Contents

WITHDRAWN

D1354500

Note to Reader

As textbooks become more expensive, authors are often asked to reduce the number of worked and unworked problems, examples and case studies. This may reduce costs, but it can be at the expense of practical work which gives point to the theory.

Checkbooks if anything lean the other way. They let problem-solving establish and exemplify the theory contained in technician syllabuses. The Checkbook reader can gain *real* understanding through seeing problems solved and through solving problems himself.

Checkbooks do not supplant fuller textbooks, but rather supplement them with an alternative emphasis and an ample provision of worked and unworked problems. The brief outline of essential data—definitions, formulae, laws, regulations, codes of practice, standards, conventions, procedures, etc—will be a useful introduction to a course and a valuable aid to revision. Short-answer and multi-choice problems are a valuable feature of many Checkbooks, together with conventional problems and answers.

Checkbook authors are carefully selected. Most are experienced and successful technical writers; all are experts in their own subjects; but a more important qualification still is their ability to demonstrate and teach the solution of problems in their particular branch of technology, mathematics or science.

Authors, General Editors and Publishers are partners in this major low-priced series whose essence is captured by the Checkbook symbol of a question or problem 'checked' by a tick for correct solution.

Preface

This book is the second in a series of five volumes and has been prepared to cover the standard unit of Construction Technology level 2 as set out in the guidelines produced by the Technician Education Council. It is not a comprehensive textbook or reference volume but a series of illustrations giving the basic notes and details required by the students at this level.

The book has been prepared on the assumption that the reader has completed a course of study in Construction Technology at level 1 and therefore has a basic knowledge of building construction and technology. In any consecutive set of syllabuses there is always a certain amount of repetition and reiteration within the syllabus content and as far as possible this book has been prepared without repeating the details and data given in the first volume covering the level 1 syllabus. Students may therefore find it necessary to refer back to the first volume in order to revise some of the basic principles to achieve full comprehension of the extension of some topic areas.

The series of questions given are typical of those which may be encountered in phase tests and assignments. Since there may be a number of correct answers to some of the questions formal answers have not been given, but the student is referred to the illustration(s) in which the relevant information can be found so that the answers may be checked by the students themselves.

To fully comprehend the subject of construction technology the reader is recommended to seek out all other sources of information such as that found in textbooks, works of reference, British Standards, Building Research Establishment Digests, Building Regulations and manufacturers' literature, but above all to observe and study buildings and building works so that the theory learnt can be related to practical examples.

R Chudley
Guildford County College of Technology

Butterworths Technical and Scientific Checkbooks

General Editor for Building, Civil Engineering, Surveying and Architectural titles:
Colin R. Bassett, lately of Guildford County College of Technology.

General Editors for Science, Engineering and Mathematics titles:
J.O. Bird and A.J.C. May, Highbury College of Technology, Portsmouth.

A comprehensive range of Checkbooks will be available to cover the major syllabus areas of the TEC, SCOTEC and similar examining authorities. A comprehensive list is given below and classified according to levels.

Level 1 (Red covers)
Mathematics
Physical Science
Physics
Construction Drawing
Construction Technology
Microelectronic Systems
Engineering Drawing
Workshop Processes & Materials

Level 2 (Blue covers)
Mathematics
Chemistry
Physics
Building Science and Materials
Construction Technology
Electrical & Electronic Applications
Electrical & Electronic Principles
Electronics
Microelectronic Systems
Engineering Drawing
Engineering Science
Manufacturing Technology

Level 3 (Yellow covers)
Mathematics
Chemistry
Building Measurement
Construction Technology
Environmental Science
Electrical Principles
Electronics
Electrical Science
Mechanical Science
Engineering Mathematics & Science

Level 4 (Green covers)
Mathematics
Building Law
Building Services & Equipment
Construction Site Studies
Concrete Technology
Economics of the Construction Industry
Geotechnics
Engineering Instrumentation & Control

Level 5
Building Services & Equipment
Construction Technology

1 General

To check answers to the following questions the student should refer to the information given in the figure number(s) quoted at the end of each question.

SHORT ANSWER QUESTIONS

1 What is the basic objective of site investigations for new works?   [*Fig 1*]

2 What is the purpose of trial pits and hand auger holes in the context of soil investigations?   [*Fig 2*]

3 Give typical dimensions for trial pits and hand auger holes used in soil investigations.   [*Fig 2*]

4 Under which Acts of Parliament can local authorities make tree preservation orders?   [*Fig 3*]

5 Trees on building sites which are covered by a tree preservation order should be protected by a suitable fence. Describe or sketch a fence suitable for this purpose.   [*Fig 3*]

6 What are the three primary objectives of site security?   [*Fig 4*]

7 In the context of temporary works what is a hoarding?   [*Fig 5*]

8 Assuming a minimum area of 3.7 m² per person and a minimum volume of 11.5 m³ per person calculate a suitable size of office for a site agent plus an allowance for 3 visitors if the office is to be constructed to a module dimension of 300 mm.   [*Fig 7*]

9 Name the Statutory Instrument which covers the health and welfare provisions for persons on construction sites.   [*Fig 8*]

10 List five fence types which would be suitable for use as a protection fence around a site storage compound.   [*Fig 9*]

11 By means of a neat sketch show how loose building blocks should be stored on an open site.   [*Fig 10*]

12 How should corrugated or similar sheet materials be stored on site?   [*Fig 11*]

13 Which authority is responsible for the following services encountered on a construction site?
(a) Electricity transmission lines; (b) Telephone cables; (c) Gas mains.   [*Fig 12*]

LONG ANSWER QUESTIONS

1 List or show by means of a neat diagram the typical data required when carrying out site investigations for new works.   [*Fig 1*]

2 To obtain soil samples for ascertaining their properties trial pits and hand auger holes can be used. By means of description and neat sketches compare these two methods stating clearly their respective advantages.   [*Fig 2*]

3 Trees and certain buildings can be considered as part of our national heritage and can be protected by law. Comment on this statement and give details of the legal protection which can be given.   [*Fig 3*]

4 List or show by means of a neat diagram typical site security provisions.   [*Fig 4*]

5 Draw a fully dimensioned and annotated detail of a typical timber hoarding with a built-up walkway which encroaches onto the highway.   [*Figs 5 and 6*]

6 The type of office accommodation provided for site staff is a matter of individual choice. Comment on this statement and give two different examples of typical site office accommodation units.   [*Fig 7*]

7 Using a copy of the Construction (Health and Welfare) Regulations 1966 as a reference give the specific minimum requirements for meals rooms, washing and sanitary facilities required on construction sites.   [*Fig 8*]

8 Materials stored on site may require protection for security reasons. Comment on this statement and sketch typical details for a storage compound fence using either a close boarded or chain link fence.   [*Figs 5, 6 and 9*]

9 Bricks, blocks and tiles may be supplied loose or strapped in unit loads. Describe or illustrate typical storage provisions for these materials on an open site.   [*Fig 10*]

10 Drainage pipes can be supplied loose and may have either spigot and socket joints or be straight barrel in format. Describe and/or sketch how each type could be stored on site.   [*Fig 11*]

11 Describe three methods which could be used to locate existing services on a building site.   [*Fig 12*]

MULTI-CHOICE QUESTIONS

1 Spot levels are:
(a) type of theodolite;   (c) level at a definite point;
(b) bench marks;   (d) contour line level.   [*Fig 1*]

2 A trial pit should provide sufficient access space for operatives which can be achieved with a minimum plan size of:
(a) 120 × 120 mm;   (c) 2100 × 2100 mm;
(b) 1200 × 1200 mm;   (d) 200 × 200 mm.   [*Fig 2*]

3 A protective fence around a tree covered by a preservation order should have a height of at least:
(a) 1200 mm; (b) 900 mm; (c) 600 mm; (d) 1000 mm. [*Fig 3*]

4 A perimeter fence around a building site to provide site security should have a height of at least:
(a) 900 mm; (b) 1000 mm; (c) 1200 mm; (d) 1800 mm. [*Fig 4*]

5 Before a hoarding can be erected a licence or permit must be obtained from the local authority who will usually require notice of:
(a) 10 to 20 hours;   (c) 3 to 7 days;
(b) 10 to 20 days;   (d) 24 hours. [*Fig 5*]

6 The clear space in front of a hoarding should be at least:
(a) 600 mm; (b) 900 mm; (c) 1000 mm; (d) 1200 mm. [*Fig 6*]

7 On construction sites a first aid box must be provided when the number of persons on site is:
(a) 0 to 3,   (c) more than 5;
(b) 3 to 5;   (d) no actual requirement. [*Fig 8*]

8 One of the main advantages of a chain link fence as a protective screen around a storage compound is that it:
(a) gives visual security;   (c) does not require painting;
(b) easy to climb;   (d) can be electrified. [*Fig 9*]

9 Ridge tiles should be stored:
(a) in cardboard boxes;   (c) lengthwise on their back;
(b) lengthwise on their bedding edges; (d) on their ends. [*Fig 10*]

10 The recommended maximum height for a stack of loose bricks is:
(a) 1200 mm; (b) 2400 mm; (c) 3600 mm; (d) no limit. [*Fig 10*]

11 Timber and joinery items should be stored:
(a) vertically and covered;
(b) vertically and uncovered;
(c) horizontally and covered with provision for free air flow;
(d) horizontally and uncovered. [*Fig 11*]

12 Public sewers are the responsibility of:
(a) Gas Board,   (c) local authority;
(b) drainage company;   (d) owner of land under which sewer passes.
[*Fig 12*]

13 On a concrete service marker the top figure on the face plate indicates:
(a) diameter in inches;   (c) distance away in feet;
(b) diameter in millimetres;   (d) distance away in metres. [*Fig 12*]

3

Site Investigation For New Works ~ the basic objective of this form of site investigation is to collect systematically and record all the necessary data which will be needed or will help in the design and construction processes of the proposed work. The collected data should be presented in the form of fully annotated and dimensioned plans and sections. Anything on adjacent sites which may affect the proposed works or conversely anything appertaining to the proposed works which may affect an adjacent site should also be recorded.

Typical Data Required ~

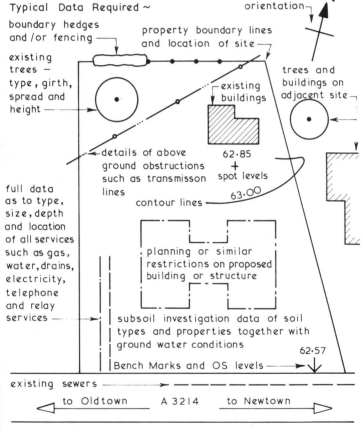

Fig 1   Site investigations

4

Purpose ~ primarily to obtain subsoil samples for identification, classification and ascertaining the subsoil's characteristics and properties. Trial pits and augered holes may also be used to establish the presence of any geological faults and the upper or lower limits of the water table.

Typical Details ~

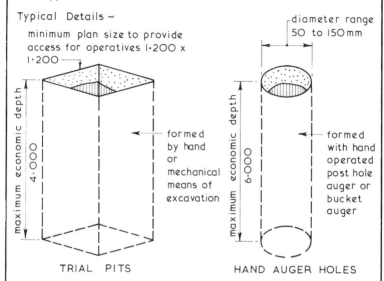

minimum plan size to provide access for operatives 1·200 x 1·200

maximum economic depth 4·000

formed by hand or mechanical means of excavation

TRIAL PITS

diameter range 50 to 150mm

maximum economic depth 6·000

formed with hand operated post hole auger or bucket auger

HAND AUGER HOLES

General use ~

dry ground which requires little or no temporary support to sides of excavation.

Subsidiary use ~
to expose and /or locate underground services.

Advantages ~
subsoil can be visually examined insitu — both disturbed and undisturbed samples can be obtained.

General use ~

dry ground but liner tubes could be used if required to extract subsoil samples at a depth beyond the economic limit of trial holes.

Advantages ~
generally a cheaper and simpler method of obtaining subsoil samples than the trial pit method.

Trial pits and holes should be sited so that the subsoil samples will be representative but not interfering with works

Fig 2 Trial pits and hand auger holes

Trees~ these are part of our national heritage and are also the source of timber – to maintain this source a control over tree felling has been established under the Forestry Act 1967 which places the control responsibility on the Forestry Commisson. Local planning authorities also have powers under the Town and Country Planning Act 1971 and the Town and Country Amenities Act 1974 to protect trees by making tree preservation orders. Contravention of such an order can lead to a substantial fine and a compulsion to replace any protected tree which has been removed or destroyed. Trees on building sites which are covered by a tree preservation order should be protected by a suitable fence.

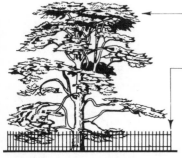

←——— tree covered by a tree preservation order

— cleft chestnut or similar fencing at least 1·200 high erected to the full spread and completely encircling the tree

Trees, shrubs, bushes and tree roots which are to be removed from site can usually be grubbed out using hand held tools such as saws, picks and spades. Where whole trees are to be removed for relocation special labour and equipment is required to ensure that the roots, root earth ball and bark are not damaged.

Structures ~ buildings which are considered to be of historic or architectural interest can be protected under the Town and Country Acts provisions. The Department of the Environment lists buildings according to age, architectural, historical and / or intrinsic value. It is an offence to demolish or alter a listed building without first obtaining 'listed building consent' from the local planning authority. Contravention is punishable by a fine and / or imprisonment. It is also an offence to demolish a listed building without giving notice to the Royal Commission on Historic Monuments, this is to enable them to note and record details of the building.

Fig 3   Protection orders for trees and structures

Site Security ~ the primary objectives of site security are-
1. Security against theft.
2. Security from vandals.
3. Protection from innocent trespassers.

The need for and type of security required will vary from site to site according to the neighbourhood, local vandalism record and the value of goods stored on site. Perimeter fencing, internal site protection and night security may all be necessary.

Typical Site Security Provisions ~

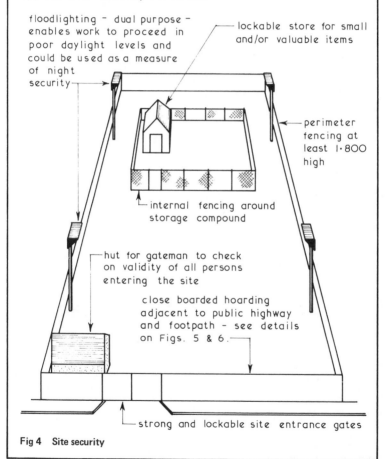

floodlighting – dual purpose – enables work to proceed in poor daylight levels and could be used as a measure of night security

lockable store for small and/or valuable items

perimeter fencing at least 1·800 high

internal fencing around storage compound

hut for gateman to check on validity of all persons entering the site

close boarded hoarding adjacent to public highway and footpath – see details on Figs. 5 & 6.

strong and lockable site entrance gates

Fig 4 Site security

Hoardings ~ under the Highways Act 1959 a close boarded fence hoarding must be erected prior to the commencement of building operations if such operations are adjacent to a public footpath or highway. The hoarding needs to be adequately constructed to provide protection for the public, resist impact damage, resist anticipated wind pressures and adequately lit at night. Before a hoarding can be erected a licence or permit must be obtained from the local authority who will usually require 10 to 20 days notice. The licence will set out the minimum local authority requirements for hoardings and define the time limit period of the licence.

Typical Hoarding Details ~

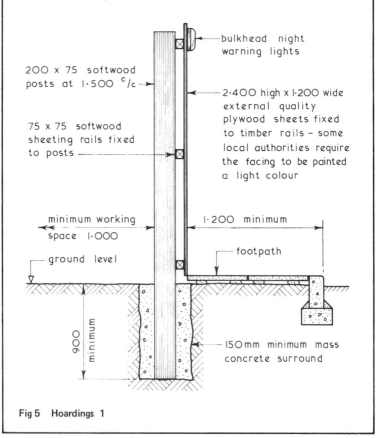

200 x 75 softwood posts at 1·500 ^c/c →

75 x 75 softwood sheeting rails fixed to posts —

minimum working space 1·000

ground level

900 minimum

bulkhead night warning lights

2·400 high x 1·200 wide external quality plywood sheets fixed to timber rails - some local authorities require the facing to be painted a light colour

1·200 minimum

footpath

150mm minimum mass concrete surround

Fig 5   Hoardings 1

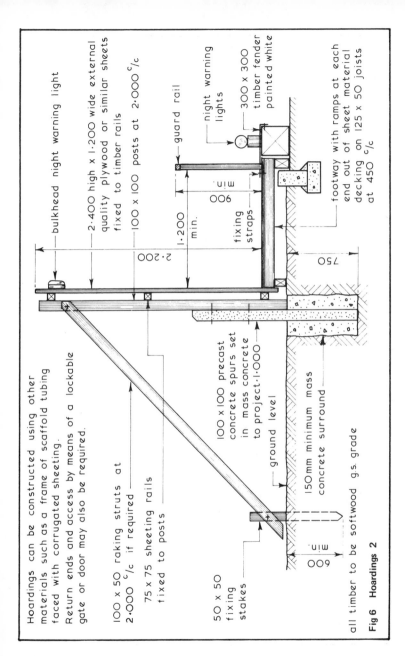

Hoardings can be constructed using other materials such as a frame of scaffold tubing faced with corrugated sheeting.
Return ends and access by means of a lockable gate or door may also be required.

100 x 50 raking struts at 2·000 c/c if required

75 x 75 sheeting rails fixed to posts

50 x 50 fixing stakes

100 x 100 precast concrete spurs set in mass concrete to project 1·000

ground level

150mm minimum mass concrete surround

all timber to be softwood g.s. grade

600 min

bulkhead night warning light

2·400 high x 1·200 wide external quality plywood or similar sheets fixed to timber rails

100 x 100 posts at 2·000 c/c

guard rail

night warning lights

300 x 300 timber fender painted white

footway with ramps at each end out of sheet material decking on 125 x 50 joists at 450 c/c

2·200

1·200 min.

900 min

fixing straps

750

Fig 6 Hoardings 2

9

Office Accommodation ~ the type of office accommodation to be provided on site is a matter of choice for each individual contractor who can use timber framed huts, prefabricated cabins, mobile offices or even caravans. Generally separate offices would be provided for site agent, clerk of works, administrative staff and site surveyors.

The minimum requirements of such accommodation is governed by the Offices, Shops and Railway Premises Act 1963 unless they are ~

1. Mobile units in use for not more than 6 months.
2. Fixed units in use for not more than 6 weeks.
3 Any type of unit in use for not more than 21 man hours per week.
4. Office for exclusive use of self employed person.
5. Office used by family only staff.

Sizing Example ~

Office for site agent and assistant plus an allowance for 3 visitors.

Assume an internal average height of 2·400.
Allow $3·7$ m^2 minimum per person and $11·5$ m^3 minimum per person.

Minimum area $= 5 \times 3·7 = 18·5$ m^2

Minimum volume $= 5 \times 11·5 = 57·5$ m^3

Assume office width of 3·000 then minimum length required

is $= \dfrac{57·5}{3 \times 2·4} = \dfrac{57·5}{7·2} = 7·986$ say 8·000

Area check $3 \times 8 = 24$ m^2 which is $> 18·5$ m^2 ∴ satisfactory

Typical Examples ~

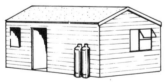

Timber framed site hut with insulated walls and roof. Wall cladding - painted weatherboard. Hut supplied unequipped. Sizes based on 1·500 module.

Portable cabin with four steel tube jacking legs. Shell-timber framing with plywood facing. Cabin insulated and fully equipped. Wide range of sizes available.

Fig 7   Site office accommodation

The minimum requirements for health and welfare provisions for persons on construction sites are set out in the Construction (Health and Welfare) Regulations 1966 and the Construction (Health and Welfare)(Amended) Regulations 1974.

Provision	Requirement	No. of persons employed on site
FIRST AID	Box to be distinctively marked and in charge of responsible person	5 to 50 - first aid boxes 50 + first aid box and a person trained in first aid
AMBULANCES	Stretcher(s) in charge of responsible person	25 + notify ambulance authority of site details within 24 hours of employing more than 25 persons
FIRST AID ROOM	Used only for rest or treatment and in charge of trained person	If more than 250 persons employed on site each employer of more than 40 persons to provide a first aid room
SHELTER AND ACCOMMODATION FOR CLOTHING	All persons on site to have shelter and a place for depositing clothes	Up to 5 where possible a means of warming themselves and drying wet clothes 5 + adequate means of warming themselves and drying wet clothing
MEALS ROOM	Drinking water, means of boiling water and eating meals for all persons on site	10 + facilities for heating food if hot meals are not available on site
WASHING FACILITIES	Washing facilities to be provided for all persons on site for more than 4 hours	20 to 100 if work is to last more than 6 weeks - hot and cold or warm water, soap and towel 100 + work lasting more than 12 months - 4 wash places + 1 for every 35 persons over 100
SANITARY FACILITIES	To be maintained, lit and kept clean. Separate facilities for female staff	Up to 100 - 1 convenience for every 25 persons 100 + - 1 convenience for every 35 persons

Fig 8   Site health and welfare requirements

Site Storage ~ materials stored on site prior to being used or fixed may require protection for security reasons or against the adverse effects which can be caused by exposure to the elements.

Small and Valuable Items ~ these should be kept in a secure and lockable store. Similar items should be stored together in a rack or bin system and only issued against an authorised requisition.

Large or Bulk Storage Items ~ for security protection these items can be stored within a lockable fenced compound. The form of fencing chosen may give visual security by being of an open nature but these are generally easier to climb than the close boarded type of fence which lacks the visual security property.

Typical Storage Compound Fencing ~

Close boarded fences can be constructed on the same methods used for hoardings - see Figs. 5 & 6.

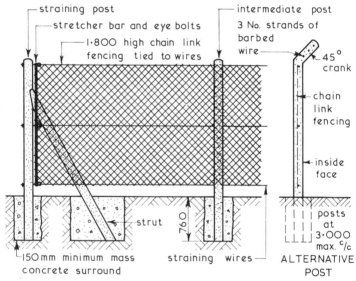

CHAIN LINK FENCING WITH PRECAST CONCRETE POSTS

Alternative Fence Types ~ woven wire fence, strained wire fence, cleft chestnut pale fence, wooden palisade fence, wooden post and rail fence and metal fences - see BS1722 for details.

Fig 9   Site storage 1

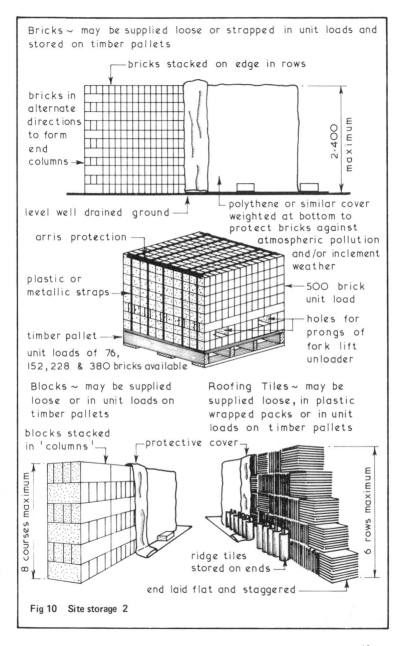

Bricks ~ may be supplied loose or strapped in unit loads and stored on timber pallets

bricks stacked on edge in rows

bricks in alternate directions to form end columns →

2·400 maximum

level well drained ground ─┘

polythene or similar cover weighted at bottom to protect bricks against atmospheric pollution and/or inclement weather

arris protection ─┐

plastic or metallic straps ─

500 brick unit load

holes for prongs of fork lift unloader

timber pallet ─
unit loads of 76, 152, 228 & 380 bricks available

Blocks ~ may be supplied loose or in unit loads on timber pallets

blocks stacked in 'columns'

8 courses maximum

Roofing Tiles ~ may be supplied loose, in plastic wrapped packs or in unit loads on timber pallets

protective cover

6 rows maximum

ridge tiles stored on ends ─

end laid flat and staggered ─

Fig 10   Site storage 2

13

Drainage Pipes~ supplied loose or strapped together on timber pallets

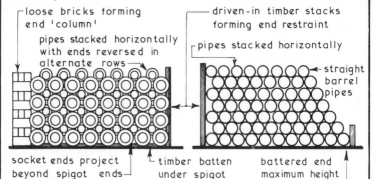

loose bricks forming end 'column'

pipes stacked horizontally with ends reversed in alternate rows

driven-in timber stacks forming end restraint

pipes stacked horizontally

straight barrel pipes

socket ends project beyond spigot ends

timber batten under spigot

battered end maximum height 1·500

gullies etc., should be stored upside down and supported to remain level

Baths~ stacked or nested vertically or horizontally on timber battens

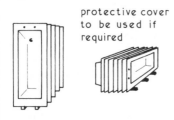

protective cover to be used if required

Timber and Joinery Items~ should be stored horizontally and covered but with provison for free air flow

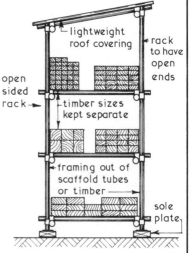

lightweight roof covering

rack to have open ends

open sided rack

timber sizes kept separate

framing out of scaffold tubes or timber

sole plate

Basins~ stored similar to baths but not more than four high if nested one on top of another

Corrugated and Similar Sheet Materials~ stored flat on a level surface and covered with a protective polythene or similar sheet material

Cement, Sand and Aggregates ~ for supply and storage details see Figs. 22 & 23.

Fig 11   Site storage 3

Services which may be encountered on construction sites and the authority responsible are ~

Water — Local Water Company

Electricity – transmission ~ Central Electricity Generating Board in England and Wales.

distribution ~ Area Electricity Board in England and Wales.

South of Scotland Electricity Board and the North of Scotland Hydro-Electric Board.

Gas – Area Gas Board.

Telephones ~ British Telecom.

Drainage – Local Authority unless a private drain or sewer when owner(s) is responsible.

All the above authorities must be notified of any proposed new services and alterations or terminations to existing services before any work is carried out.

Locating Existing Services on Site ~

Method 1 – By reference to maps and plans prepared and issued by the respective responsible authority

Method 2 – Using visual indicators ~

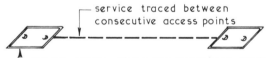

service traced between consecutive access points

access covers which are sometimes marked as to the service below e.g. soil or surface water

precast concrete cover

diameter (inches)
distance (feet)

SV
4
.10.

concrete markers indicating diameter of service and the distance in front of the marker

unearthed protective covers

electric cable

Method 3 – Detection-specialist contractor employed to trace all forms of underground service using electronic subsurface survey equipment.

Once located position and type of service can be plotted on a map or plan, marked with special paint on hard surfaces and marked with wood pegs with identification data on earth surfaces.

Fig 12   Locating public utility services

2 Substructure

To check answers to the following questions the student should refer to the information given in the figure number(s) quoted at the end of each question.

SHORT ANSWER QUESTIONS

1 For the removal of top soil over the site of a building what types of plant could be employed? [*Fig 13*]

2 Name the Statutory Instrument in which the minimum legal requirements regarding the support and protection of excavations are given. [*Fig 14*]

3 State the depth at which an excavation must have a protective barrier or be covered and give the depth at which a suitable supply of timber or other suitable material must be available to provide temporary support if required. [*Fig 15*]

4 In the context of support work to excavations define the terms 'poling board' and 'trench sheeting.' [*Fig 16*]

5 By example show how the width of a mass concrete strip foundation may be calculated. [*Fig 17*]

6 By comparison show the difference in the use of level and stepped foundations on a sloping site. [*Fig 18*]

7 Give two methods of increasing the resistance of a mass concrete strip foundation against induced tensile stresses. [*Fig 19*]

8 Give two situations in which short bored piles could be considered as an alternative to traditional mass concrete strip foundations. [*Fig 20*]

9 In a simple reinforced concrete raft foundation with an edge thickening what is the purpose of the permanent concrete perimeter paving? [*Fig 21*]

10 Name all the materials which are used in the production of a concrete mix. [*Fig 22*]

11 Name four materials which could be used as a lightweight aggregate in a concrete mix. [*Fig 22*]

12 Define the basic requirements for the storage of cement on a construction site.
[*Fig 23*]

13 Concrete can be batched by volume or weight. Briefly compare these two methods.
[*Figs 24 and 25*]

14 What are the functions of the water in a concrete mix? [*Fig 25*]

15 What are the main advantages of using ready mixed concrete? [*Fig 26*]

16 State the sequence of operations for the erection and striking of formwork for a simple concrete beam. [*Fig 28*]

LONG ANSWER QUESTIONS

1 Site clearance, removal of top soil and excavations up to 2.500 deep can be carried out by manual or mechanical means. Comment on this statement and suggest a suitable method for excavating a drainage trench with an average depth of 1.500.
[*Fig 13*]

2 What are the major factors to be considered when deciding on a form of temporary support for excavations? [*Fig 14*]

3 Draw a fully dimensioned and annotated detail of a suitable temporary timber support system for a trench excavation 1.200 wide × 1.500 deep in a firm subsoil.
[*Fig 15*]

4 Draw a fully dimensioned and annotated detail of a suitable temporary timber support system for a trench excavation 1.200 wide × 2.000 deep in a wet subsoil.
[*Fig 16*]

5 A column is to be built off a square concrete foundation, if the column load is 450 kN and the bearing capacity of the subsoil is 150 kN/m^2 what is the minimum size of base which could be used? To a suitable scale draw a section through a typical reinforced concrete foundation suitable for the above condition.
[*Figs 17 and 19*]

6 Sketch the detail of a typical stepped strip foundation constructed in mass concrete and show the minimum requirements of the overlap. [*Fig 18*]

7 By means of annotated diagrams show the affect of excessive tensile stress in a mass concrete foundation and how this can be overcome by the introduction of steel reinforcement. [*Fig 19*]

8 Draw a fully dimensioned and annotated detail of a short bored pile foundation supporting a cavity wall. [*Fig 20*]

9 Simple reinforced concrete foundations can be designed with either edge thickening or an edge beam. Draw a fully dimensioned and annotated detail through one of these forms. [*Fig 21*]

10 Concrete is a mixture of four materials. Describe in detail these materials in the context of concrete production. [*Fig 22*]

11 Describe and/or sketch typical site storage methods for cement and aggregates.
[*Figs 22 and 23*]

12 By means of a neat sketch show a typical gauge box for the volume batching of concrete. [*Fig 24*]

13 When loading a weighing hopper for the weight batching of concrete it should be loaded in a specific order. Comment on this statement and show how the loading dial can be set to ensure consistent mixes. [*Fig 25*]

14 The supply of concrete is usually geared to the demand or the rate at which it can be placed. Comment on this statement in the context of small, medium and large batch requirements. [*Fig 26*]

15 Describe and illustrate as necessary the basic principles of formwork. [*Fig 27*]

16 By means of a fully dimensioned and annotated detail show the formwork for a concrete beam 300 mm wide × 600 mm deep where the soffit level is 2.400 above the ground. [*Fig 28*]

MULTI-CHOICE QUESTIONS

1 Some methods other than by manual means of removing spoil from an excavation will have to be employed if the excavation depth exceeds:
(a) 900 mm; (b) 1000 mm; (c) 1200 mm; (d) 1500 mm. [*Fig 13*]

2 The angle of repose of dry clay is:
(a) greater than wet clay;   (c) same as wet clay;
(b) less than wet clay;   (d) 0°. [*Fig 14*]

3 Spoil heap barriers to an excavation over 1.980 deep should have a minimum height of:
(a) 450 mm; (b) 600 mm; (c) 900 mm; (d) 1000 mm. [*Fig 15*]

4 In temporary support work to excavations a runner is:
(a) a poling board;   (c) vertically driven support board;
(b) another name for a waling;   (d) horizontally driven support board. [*Fig 16*]

5 The projection of a strip foundation measured from the supported wall should be:
(a) not less than wall thickness;
(b) not less than half foundation width;
(c) not less than 250 mm;
(d) not less than foundation thickness or 150 mm minimum.
[*Fig 17*]

6 The overlap of a stepped mass concrete strip foundation should be:
(a) not less than depth of strip foundation with a minimum of 300 mm;
(b) twice depth of strip foundation;
(c) 600 mm minimum;
(d) as (a) but with a maximum of 300 mm. [*Fig 18*]

7 Main reinforcement bars in a strip foundation should be placed at:
(a) right angles to length;   (c) both directions;
(b) longitudinally;   (d) only distribution bars required. [*Fig 19*]

8 The usual diameter range for short bored piles is:
(a) 100–200 mm;   (c) 250–300 mm;
(b) 150–250 mm;   (d) 350–450 mm. [*Fig 20*]

9 The usual minimum specified depth for an edge beam to a simple reinforced concrete raft foundation is:
(a) 300 mm; (b) 600 mm; (c) 750 mm; (d) 1000 mm. [*Fig 21*]

10 Coarse aggregates are those which are retained on a sieve with a mesh size of:
   (a) 5 mm; (b) 10 mm; (c) 15 mm; (d) 20 mm. [*Fig 22*]

11 A standard bag of cement contains:
   (a) 25 kg; (b) 30 kg; (c) 45 kg; (d) 50 kg. [*Fig 22*]

12 As a general rule the allowance for the bulking of damp sand is:
   (a) 10%; (b) 25%; (c) 35%; (d) damp sand does not bulk. [*Fig 24*]

13 The actual water cement ratio required to set cement is approximately:
   (a) 0.1; (b) 0.2; (c) 0.3; (d) 0.4. [*Fig 25*]

14 Small concrete batch mixers usually have an output capacity range of up to:
   (a) 50 litres;   (c) 200 litres;
   (b) 150 litres;   (d) 250 litres. [*Fig 26*]

15 The transverse soffit support member for simple beam formwork is called:
   (a) headtree; (b) prop; (c) cleat; (d) strut. [*Fig 28*]

Site Clearance and Removal of Top Soil ~

On small sites this could be carried out by manual means using hand held tools such as picks, shovels and wheelbarrows.

On all sites mechanical methods could be used the actual plant employed being dependent on factors such as volume of soil involved, nature of site and time elements.

top soil - upper level of earth usually not exceeding 300mm deep

excavation- mechanical shovel or bulldozer

lorry or dumper to transport spoil

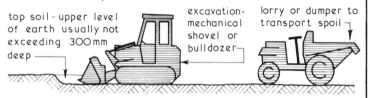

Reduced Level Excavations ~

On small sites - hand processes as given above

On all sites mechanical methods could be used dependent on factors given above.

bulldozer for cut and fill operations

mechanical shovel and attendant lorries for cut only operations

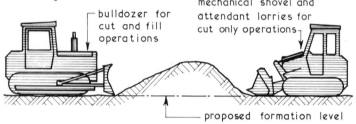

proposed formation level

Trench and Pit Excavations ~

On small sites - hand processes as given above but if depth of excavation exceeds 1·200 some method of removing spoil from the excavation will have to be employed.

On all sites mechanical methods could be used dependent on factors given above.

on large sites a trenching machine could be used

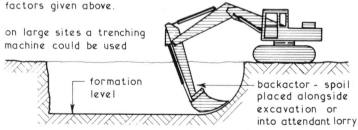

formation level

backactor - spoil placed alongside excavation or into attendant lorry

Fig 13   Excavations up to 2.500 deep — processes

All subsoils have different abilities in remaining stable during excavation works. Most will assume a natural angle of repose or rest unless given temporary support. The presence of ground water apart from creating difficult working conditions can have an adverse effect on the subsoil's natural angle of repose.

Typical Angles of Repose ~

Excavations cut to a natural angle of repose are called battered.

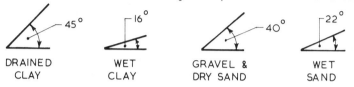

DRAINED CLAY — 45°   WET CLAY — 16°   GRAVEL & DRY SAND — 40°   WET SAND — 22°

Factors for Temporary Support of Excavations ~

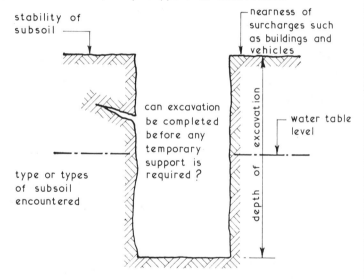

stability of subsoil

nearness of surcharges such as buildings and vehicles

can excavation be completed before any temporary support is required ?

water table level

type or types of subsoil encountered

depth of excavation

Time factors such as period during which excavation will remain open and the time of year when work is carried out.

The minimum legal requirements regarding the support and protection of excavations are set out in the Construction (General Provisions) Regulations 1961.

Fig 14   Excavations up to 2.500 deep — temporary support 1

21

Temporary Support ~ in the context of excavations this is called timbering irrespective of the actual materials used. If the sides of the excavation are completely covered with timbering it is known as close timbering whereas any form of partial covering is called open timbering.

An adequate supply of timber or other suitable material must be available and used to prevent danger to any person employed in an excavation over 1·210 deep from a fall or dislodgement of materials forming the sides of an excavation.

A suitable barrier or fence must be provided to the sides of all excavations over 1·980 deep or alternatively they must be securely covered.

Materials must not be placed near to the edge of any excavation, nor must plant be placed or moved near to any excavation so that persons employed in the excavation are endangered.

Typical Example ~

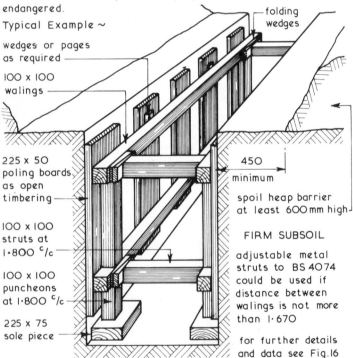

folding wedges

wedges or pages as required

100 x 100 walings

225 x 50 poling boards as open timbering

100 x 100 struts at 1·800 ᶜ/c

100 x 100 puncheons at 1·800 ᶜ/c

225 x 75 sole piece

450 minimum

spoil heap barrier at least 600 mm high

FIRM SUBSOIL

adjustable metal struts to BS 4074 could be used if distance between walings is not more than 1·670

for further details and data see Fig.16

Fig 15   Excavations up to 2.500 deep — temporary support 2

Poling Boards ~ a form of temporary support which is placed in position against the sides of excavation after the excavation work has been carried out. Poling boards are placed at centres according to the stability of the subsoils encountered.

Runners ~ a form of temporary support which is driven into position ahead of the excavation work either to the full depth or by a drive and dig technique where the depth of the runner is always lower than that of the excavation.

Trench Sheeting ~ form of runner made from sheet steel with a trough profile – can be obtained with a lapped joint or an interlocking joint.

Water ~ if present or enters an excavation a pit or sump should be excavated below the formation level to act as collection point from which the water can be pumped away.

Typical Example ~

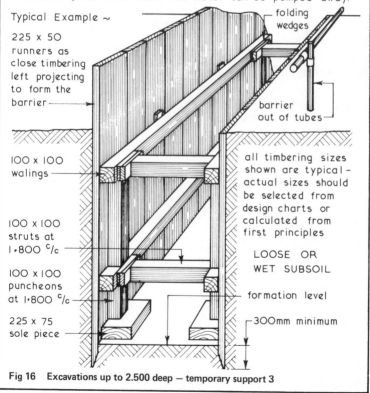

225 x 50 runners as close timbering left projecting to form the barrier

folding wedges

barrier out of tubes

100 x 100 walings

all timbering sizes shown are typical – actual sizes should be selected from design charts or calculated from first principles

LOOSE OR WET SUBSOIL

100 x 100 struts at 1·800 c/c

100 x 100 puncheons at 1·800 c/c

formation level

225 x 75 sole piece

300mm minimum

Fig 16   Excavations up to 2.500 deep — temporary support 3

Basic Sizing ~ the size of a foundation is basically dependent on two factors -
1. Load being transmitted.
2. Bearing capacity of subsoil under proposed foundation.

Bearing capacities for different types of subsoils may be obtained from tables such as than in CP 101 or from soil investigation results.

Typical Examples ~

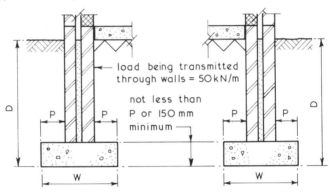

safe bearing capacity of compact gravel subsoil = 100 kN/m²

$W = \dfrac{load}{bearing\ capacity} = \dfrac{50}{100}$

$= 500\,mm$ minimum

safe bearing capacity of clay subsoil = 80 kN/m²

$W = \dfrac{load}{bearing\ capacity} = \dfrac{50}{80}$

$= 625\,mm$ minimum

The above widths may not provide adequate working space within the excavation and can be increased to give required space. Minimum width for a limited range of strip foundations can be taken direct from the table to Building Regulation D7

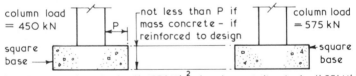

bearing capacity of subsoil 150 kN/m²

area of base $= \dfrac{load}{b.c.} = \dfrac{450}{150}$

$= 3\,m^2$ ∴ side $= \sqrt{3} = 1\cdot732$ min.

bearing capacity of subsoil 85 kN/m²

area of base $= \dfrac{load}{b.c.} = \dfrac{575}{85}$

$= 6\cdot765\,m^2$ ∴ side $= \sqrt{6\cdot765} = 2\cdot6\,m.$

Fig 17 Foundations — basic sizing

Stepped Foundations ~ these are usually considered in the context of strip foundations and are used mainly on sloping sites to reduce the amount of excavation and materials required to provide an adequate foundation.

Comparison ~

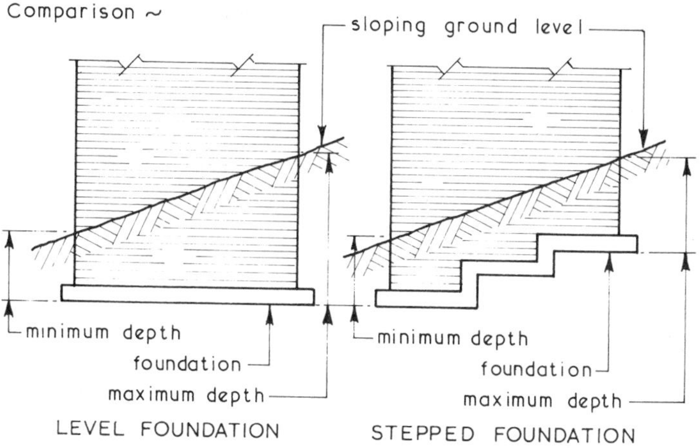

sloping ground level
minimum depth
foundation
maximum depth
LEVEL FOUNDATION
minimum depth
foundation
maximum depth
STEPPED FOUNDATION

Typical Details ~

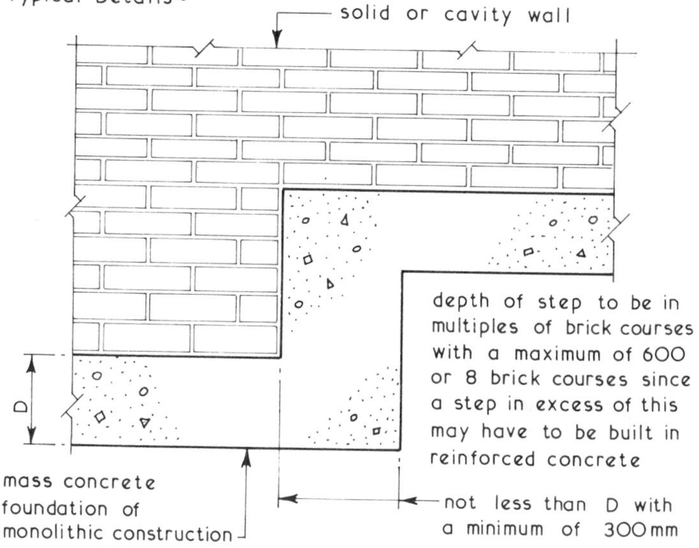

solid or cavity wall
D
depth of step to be in multiples of brick courses with a maximum of 600 or 8 brick courses since a step in excess of this may have to be built in reinforced concrete
mass concrete foundation of monolithic construction
not less than D with a minimum of 300mm

Fig 18 Stepped foundations

Concrete Foundations ~ concrete is a material which is strong in compression but weak in tension. If its tensile strength is exceeded cracks will occur resulting in a weak and unsuitable foundation. One method of providing tensile resistance is to include in the concrete foundation bars of steel as a form of reinforcement to resist all the tensile forces induced into the foundation. Steel is a material which is readily available and has high tensile strength.

Comparisons ~

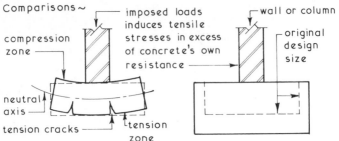

imposed loads
induces tensile
stresses in excess
of concrete's own
resistance

compression
zone

neutral
axis

tension cracks

tension
zone

wall or column

original
design
size

foundation tends to bend, the
upper fibres being compressed
and the lower fibres being
stretched and put in tension-
remedies increase size of
base or design as a reinforced
concrete foundation

size of foundation increased
to provide the resistance
against the induced tensile
stresses - generally not
economic due to the extra
excavation and materials
required

Typical RC Foundation

Reinforcement Patterns

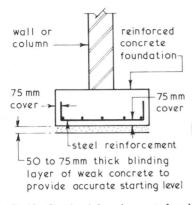

wall or
column

reinforced
concrete
foundation

75 mm
cover

75 mm
cover

steel reinforcement

50 to 75mm thick blinding
layer of weak concrete to
provide accurate starting level

distribution
bars

main bars at
right angles to
longitudinal axis

STRIP
FOUNDATION

SQUARE BASE

main bars
both ways

Fig 19   Simple reinforced concrete foundations

Short Bored Piles ~ these are a form of foundation which are suitable for domestic loadings and clay subsoils where ground movements can occur below the 900 to 1000 mm depth associated with traditional strip and trench fill foundations. They can be used where trees are planted close to the new building since the trees may eventually cause damaging ground movements due to extracting water from the subsoil and root growth. Conversely where trees have been removed this may lead to ground swelling.

Typical Details ~

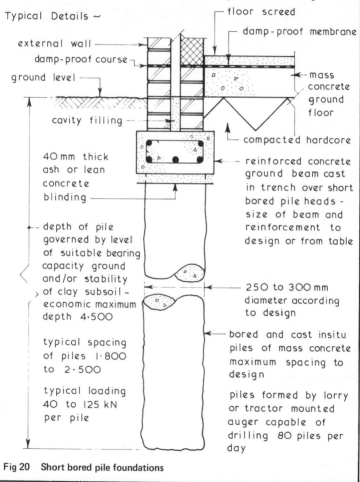

floor screed

damp-proof membrane

external wall

damp-proof course

ground level

cavity filling

40 mm thick ash or lean concrete blinding

depth of pile governed by level of suitable bearing capacity ground and/or stability of clay subsoil - economic maximum depth 4·500

typical spacing of piles 1·800 to 2·500

typical loading 40 to 125 kN per pile

mass concrete ground floor

compacted hardcore

reinforced concrete ground beam cast in trench over short bored pile heads - size of beam and reinforcement to design or from table

250 to 300 mm diameter according to design

bored and cast insitu piles of mass concrete maximum spacing to design

piles formed by lorry or tractor mounted auger capable of drilling 80 piles per day

Fig 20   Short bored pile foundations

Simple Raft Foundations ~ these can be used for lightly loaded building on poor soils or where the top 450 to 600 mm of soil is overlaying a poor quality substrata.

Typical Details ~

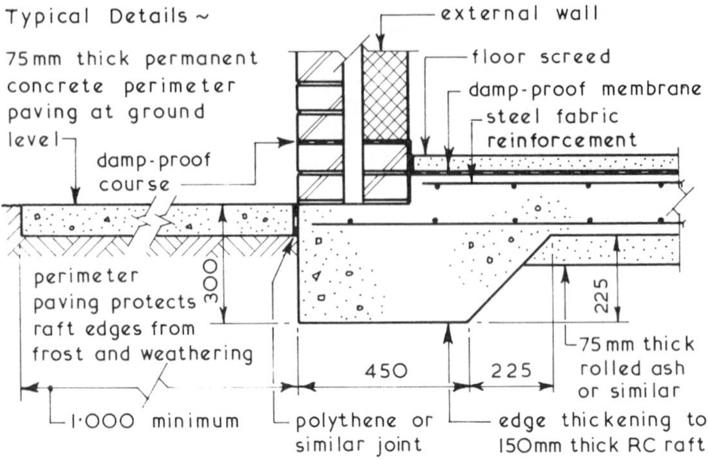

75 mm thick permanent concrete perimeter paving at ground level

damp-proof course

external wall

floor screed

damp-proof membrane

steel fabric reinforcement

perimeter paving protects raft edges from frost and weathering

300

225

75 mm thick rolled ash or similar

1·000 minimum

450

225

polythene or similar joint

edge thickening to 150 mm thick RC raft

REINFORCED CONCRETE RAFT WITH EDGE THICKENING

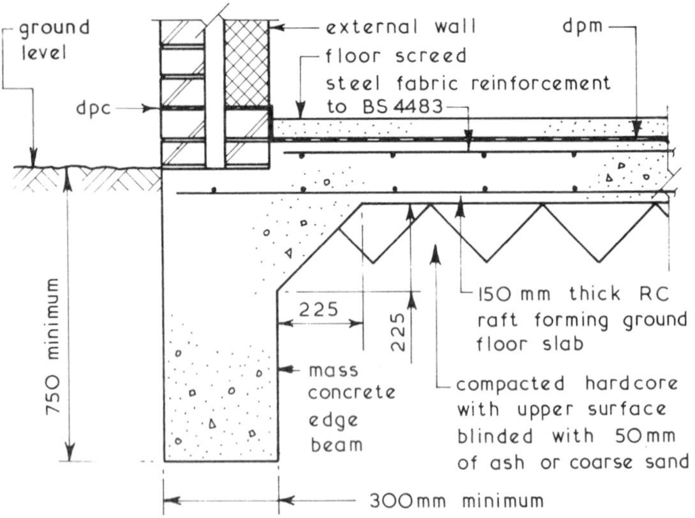

ground level

dpc

external wall

floor screed

steel fabric reinforcement to BS 4483

dpm

750 minimum

225

225

mass concrete edge beam

150 mm thick RC raft forming ground floor slab

compacted hardcore with upper surface blinded with 50 mm of ash or coarse sand

300 mm minimum

REINFORCED CONCRETE RAFT WITH EDGE BEAM

Fig 21   Simple RC raft foundations

Concrete ~ a mixture of cement + fine aggregate + coarse aggregate + water in controlled proportions and of a suitable quality.

Cement ~ powder produced from clay and chalk or limestone. In general most concrete is made with ordinary or rapid hardening Portland cement, both types being manufactured to the recommendations of BS 12. Ordinary Portland cement is adequate for most purposes but has a low resistance to attack by acids and sulphates. Rapid hardening Portland cement does not set faster than ordinary Portland cement but it does develop its working strength at a faster rate. For a concrete which must have an acceptable degree of resistance to sulphate attack sulphate resisting Portland cement made to the recommendations of BS 4027 could be specified.

50 kg

BAGS

12 t
to
50 t

SILOS

Aggregates ~ shape, surface texture and grading (distribution of particle sizes) are factors which influence the workability and strength of a concrete mix. Fine aggregates are those materials which pass through a 5mm sieve whereas coarse aggregates are those materials which are retained on a 5mm sieve. Dense aggregates are those with a density of more than $1200 \, kg/m^3$ for coarse aggregates and more than $1250 \, kg/m^3$ for fine aggregates and are covered by BS 882 – Coarse and Fine Aggregates from Natural Sources and BS 1047 Air-cooled Blastfurnace Slag Coarse Aggregates for Concrete. Lightweight aggregates include clinker – BS 1165; foamed or expanded blastfurnace slag – BS 877 and exfoliated and expanded materials such as vermiculite, perlite, clay and sintered pulverized-fuel ash to BS 3797.

coarse
aggregate

5mm sieve

fine
aggregate

Water ~ must be clean and free from impurities which are likely to affect the quality or strength of the resultant concrete. Pond, river, canal and sea water should not be used and only water fit for drinking should be specified.

drinking

Fig 22   Concrete production — materials

Cement ~ whichever type of cement is being used it must be properly stored on site to keep it in good condition. The cement must be kept dry since contact with any moisture whether direct or airborne could cause it to set. A rotational use system should be introduced to ensure that the first batch of cement delivered is the first to be used.

Typical Storage Methods ~

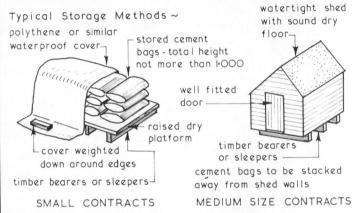

SMALL CONTRACTS   MEDIUM SIZE CONTRACTS

LARGE CONTRACTS – for bagged cement watertight shed as above for bulk delivery loose cement a cement storage silo.

Aggregates ~ essentials of storage are to keep different aggregate types and/or sizes separate, store on a clean, hard free draining surface and to keep the stored aggregates clean and free of leaves and rubbish.

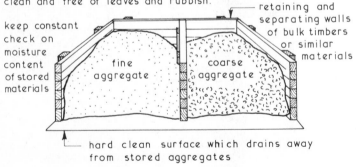

Fig 23   Concrete production — site storage of materials

Concrete Batching ~ a batch is one mixing of concrete and can be carried out by measuring the quantities of materials required by volume or weight. The main aim of both methods is to ensure that all consecutive batches are of the same standard and quality.

Volume Batching ~ concrete mixes are often quoted by ratio such as 1 : 2 : 4 (cement : fine aggregate or sand : coarse aggregate.) A bag of cement weighing 50 kg has a volume of 0·033 m^3 therefore for the above mix 2 x 0·033 (0·066 m^3) of sand and 4 x 0·033 (0·132 m^3) of coarse aggregate is required. To ensure accurate amounts of materials are used for each batch a gauge box should be employed its size being based on convenient handling. Ideally a batch of concrete should be equated to using 1 bag of cement per batch. Assuming a gauge box 300mm deep and 300 mm wide with a volume of half the required sand the gauge box size would be–

volume = length x width x depth = length x 300 x 300

$$\therefore \text{length} = \frac{\text{volume}}{\text{width x depth}} = \frac{0·033}{0·3 \text{ x } 0·3} = 0·366 \text{ m}$$

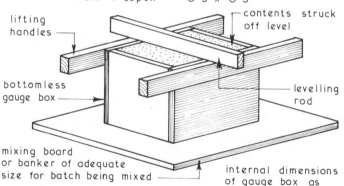

lifting handles

contents struck off level

bottomless gauge box

levelling rod

mixing board or banker of adequate size for batch being mixed

internal dimensions of gauge box as calculated

For the above given mix fill gauge box once with cement, twice with sand and four times with coarse aggregate.

An allowance must be made for the bulking of damp sand which can be as much as 33 $\frac{1}{3}$ %. General rule of thumb unless using dry sand allow for 25% bulking.

Materials should be well mixed dry before adding water.

Fig 24   Concrete production — volume batching

Weigh Batching ~ this is a more accurate method of measuring materials for concrete than volume batching since it reduces considerably the risk of variation between different batches. The weight of sand is affected very little by its dampness which in turn leads to greater accuracy in proportioning materials. When loading a weighing hopper the materials should be loaded in a specific order –

1. Coarse aggregates – tends to push other materials out and leaves the hopper clean.

2. Cement – this is sandwiched between the other materials since some of the fine cement particles could be blown away if cement is put in last.

3. Sand or Fine Aggregates – put in last to stabilise the fine lightweight particles of cement powder.

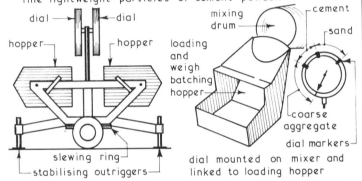

INDEPENDENT WEIGH BATCHER   INTEGRAL WEIGH BATCHER

Typical Densities ~ cement - 1440 kg/m^3 sand - 1600 kg/m^3 coarse aggregrate - 1440 kg/m^3

Water/Cement Ratio ~ water in concrete has two functions-

1. Start the chemical reaction which causes the mixture to set into a solid mass.

2. Give the mix workability so that it can be placed, tamped or vibrated into the required position.

Very little water is required to set concrete (approximately 0·2 w/c ratio) the surplus evaporates leaving minute voids therefore the more water added to the mix to increase its workability the weaker is the resultant concrete. Generally w/c ratios of 0·4 to 0·5 are adequate for most purposes.

Fig 25   Concrete production — weigh batching

Concrete Supply ~ this is usually geared to the demand or the rate at which the mixed concrete can be placed. Fresh concrete should always be used or placed within 30 minutes of mixing to prevent any undue drying out. Under no circumstances should more water be added after the initial mixing.

Small Batches ~ small easily transported mixers with

output capacities of up to 100 litres can be used for small and intermittent batches. These mixers are versatile and robust machines which can be used for mixing mortars and plasters as well as concrete.

Medium to Large Batches ~ mixers with output capacities

from 100 litres to 10 m³ with either diesel or electric motors. Many models are available with tilting or reversing drum discharge, integral weigh batching and loading hopper and a controlled water supply.

Ready Mixed Concrete ~ used mainly for large concrete
batches of up to 6 m³. This method of concrete supply has the advantages of eliminating the need for site space to accommodate storage of materials, mixing plant and the need to employ adequately trained site staff who can constantly produce reliable and consistent concrete mixes. Ready mixed concrete supply depots also have better facilities and arrangements for producing and supply mixed concrete

in winter or inclement weather conditions. In many situations it is possible to place the ready mixed concrete into the required position direct from the delivery lorry via the delivery chute or by feeding it into a concrete pump. The site must be capable of accepting the 20 tonnes laden weight of a typical ready mixed concrete lorry with a turning circle of about 15·000. The supplier will want full details of mix required and the proposed delivery schedule – see BS 5328.

Fig 26   Concrete production — supply

Basic Formwork ~ concrete when first mixed is a fluid and therefore to form any concrete member the wet concrete must be placed in a suitable mould to retain its shape, size and position as it sets. It is possible with some forms of concrete foundations to use the sides of the excavation as the mould but in most cases when casting concrete members a mould will have to be constructed on site. These moulds are usually called formwork. It is important to appreciate that the actual formwork is the reverse shape of the concrete member which is to be cast.

Basic Principles ~

formwork sides can be designed to offer all the necessary resistance to the imposed pressures as a single member or alternatively they can be designed to use a thinner material which is adequately strutted — for economic reasons the latter method is usually employed

grout tight joints

wet concrete - density is greater than that of the resultant set and dry concrete

formwork sides — limits width and shape of wet concrete and has to resist the hydrostatic pressure of the wet concrete which will diminish to zero within a matter of hours depending on setting and curing rate

formwork soffits can be designed to offer all the necessary resistance to the imposed loads as a single member or alternatively they can be designed to a thinner material which is adequately propped — for economic reasons the latter method is usually employed

formwork base or soffit – limits depth and shape of wet concrete and has to resist the initial dead load of the wet concrete and later the dead load of the dry set concrete until it has gained sufficient strength to support its own dead weight which is usually several days after casting depending on curing rate.

Fig 27 Basic formwork — principles

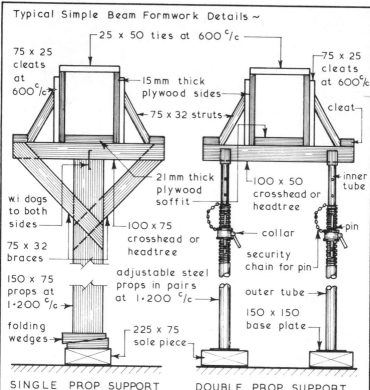

Typical Simple Beam Formwork Details ~

25 x 50 ties at 600 c/c

75 x 25 cleats at 600 c/c

15 mm thick plywood sides

75 x 32 struts

75 x 25 cleats at 600 c/c

cleat

21 mm thick plywood soffit

100 x 50 crosshead or headtree

inner tube

w.i dogs to both sides

100 x 75 crosshead or headtree

collar

pin

security chain for pin

75 x 32 braces

150 x 75 props at 1·200 c/c

adjustable steel props in pairs at 1·200 c/c

folding wedges

outer tube

150 x 150 base plate

225 x 75 sole piece

SINGLE PROP SUPPORT

DOUBLE PROP SUPPORT

Erecting Formwork

1. Props positioned and levelled through.
2. Soffit placed, levelled and position checked.
3. Side forms placed, their position checked before being fixed.
4. Strutting position and fixed.
5. Final check before casting.

Suitable Formwork Materials~ timber, steel and special plastics.

Striking or Removing Formwork

1. Side forms as soon as practicable usually within hours of casting this allows drying air movements to take place around the setting concrete.
2. Soffit formwork as soon as practicable usually within days but as a precaution some props are left in position until concrete member is self supporting. (See table in CP110.)

Fig 28   Basic formwork — details

3 Superstructure

To check answers to the following questions the student should refer to the information given in the figure number(s) quoted at the end of each question.

SHORT ANSWER QUESTIONS

1 Define and state the purposes of brick bonding. [*Fig 29*]

2 By example show the difference between broken bond and reverse bond. [*Fig 29*]

3 In solid brickwork what is English Bond? [*Fig 30*]

4 In solid brickwork what is Flemish Bond? [*Fig 31*]

5 What is Rat Trap Bond? [*Fig 32*]

6 In brick bonding what are the functions of an attached pier? [*Fig 33*]

7 Define the term parapet wall and state the basic difference between high and low parapets. [*Fig 34*]

8 State the primary functions of a damp-proof course and give the three basic ways in which it is used. [*Fig 35*]

9 List at least six different materials which could be used as a damp-proof course. [*Fig 35*]

10 State the primary functions of a support over an opening. [*Fig 36*]

11 By means of a neat sketch show the elevation of a typical segmental brick arch. [*Fig 37*]

12 Weepholes are often used above the heads of openings. State the centres at which they would be placed and state under which circumstances they would be used. [*Fig 38*]

13 By means of a neat and fully annotated sketch show the jamb detail of a cavity wall at an opening. [*Fig 39*]

14 State the primary function of an external sill at an opening in a wall. [*Fig 40*]

15 State the primary functions and basic design requirements for a pitched roof. [*Fig 41*]

16 By means of line diagrams show the difference between a lean-to roof and a couple roof. [*Fig 42*]

17 Timber pitched roofs may be terminated by an open or closed eaves. By means of an annotated sketch show the detail of one of these formats. [*Fig 43*]

18 What is the difference between a roof truss and a trussed rafter? [*Figs 44 and 45*]

19 State the main functions of a roof underlay and give two different underlay materials which could be used for a domestic pitched roof. [*Fig 46*]

20 Name the three tile formats which are usually required for general surface double lap tiling. [*Fig 47*]

21 By means of sketches show the dual purpose of a plain tile with a length of 190 mm. [*Fig 48*]

22 By means of a neat sketch show the purpose of a hip iron when used in the context of double lap tiling. [*Fig 49*]

23 What is a soaker and where would it be used? [*Fig 50*]

24 How is the gauge or batten spacing for single lap tiles determined? [*Fig 51*]

25 State how hips and valleys can be finished when covering a pitched roof with single lap tiling. [*Fig 52*]

26 Slates can be laid as head nailed or centre nailed. Show the difference in calculating the gauge for these formats. [*Fig 53*]

27 Briefly describe the two common methods of thermal insulation used in domestic pitched roofs and state any advantages and disadvantages of these methods. [*Fig 54*]

28 In the context of flat roofs what is a firring? Show by means of a neat diagram the firring arrangement for a roof where the fall is at right angles to the run of the joists. [*Fig 55*]

29 How can the size of timber roof joists for a small span flat roof be calculated? [*Fig 56*]

30 By means of neat sketches show how rainwater can be directed back onto the main roof surface at the verge and at an abutment. [*Fig 57*]

31 What is built-up roofing felt? [*Fig 58*]

32 What are the primary functions of a window? [*Fig 59*]

33 In the context of windows what is the purpose of a schedule? [*Fig 60*]

34 State the main factors to be considered when selecting glass. [*Fig 61*]

35 When securing glass with putty to timber or metal frames what are the main points to consider? [*Fig 62*]

LONG ANSWER QUESTIONS

1 By example show the difference between half, quarter and third brick bonding. [*Fig 29.*]

2 Draw the plans of alternate courses of a one brick thick wall in English Bond having a length of 2.690 with an attached pier at one end and a return wall at the other end. *[Fig 30]*

3 Draw the plans of alternate courses of a one brick thick wall in Flemish Bond having a length of 2.240 with an attached pier at one end and a return wall at the other end. *[Fig 31]*

4 Garden wall bonds can be used for solid brick walls. By example show a comparison of English and Flemish garden wall bonds and state their respective advantages. *[Fig 32]*

5 By example show how attached piers may be bonded in the length of a one brick wall and a half brick wall. *[Fig 33]*

6 Parapets may be constructed as high level or low level. Draw fully dimensioned and annotated details of a high level parapet in a cavity wall and a low level parapet in a one brick thick solid wall. *[Fig 34]*

7 By means of fully dimensioned and annotated details show how damp-proof courses may be used to prevent moisture penetration from below ground level and from horizontal entry. *[Fig 35]*

8 By comparison show how loads are transmitted around an opening by means of a beam or lintel and by means of an arch. *[Fig 36]*

9 A semi-circular arch in a one brick wall has a clear span of 1.340. Draw a fully dimensioned and annotated elevation and section of a suitable timber-framed centre assuming the springing line to be 1.500 above the ground level. *[Fig 37]*

10 Draw three fully dimensioned and annotated details through the head of an opening in a cavity wall using a different type of lintel in each case. *[Fig 38]*

11 By comparison show the difference in jamb treatments in openings in solid and cavity walls. *[Fig 39]*

12 Typical materials for external sills to openings are timber, cast stone, precast concrete and slate. Draw fully dimensioned and annotated comparative details of two of these sill types. *[Fig 40]*

13 List or show by means of a diagram the component parts of a timber pitched roof and its coverings. *[Fig 41]*

14 A timber collar roof is to be constructed over a clear span of 5.000 with a pitch of 40°. Draw a fully dimensioned and annotated detail of a typical section through the roof construction. Coverings and finishes need not be shown. *[Fig 42]*

15 In double rafter or purlin roofs the purlin can be placed vertically or at right angles to the rafters. Show by comparative details the difference between the two methods where the strut and collar arrangement occurs. *[Fig 43]*

16 By means of a fully dimensioned and annotated detail show a typical roof truss for a timber roof with a clear span of 7.000 and a pitch of 45°. *[Fig 44]*

17 By means of a fully dimensioned and annotated detail show a typical timber trussed rafter for a roof with a clear span of 6.000 and a pitch of 30°. *[Fig 45]*

18 By means of an annotated sketch show how a roof underlay should be fixed to a roof with a hip end. [*Fig 46*]

19 Briefly describe double lap tiles and tiling and by means of a dimensioned sketch show the detail of a typical plain tile. [*Fig 47*]

20 To a suitable scale draw a fully dimensioned and annotated detail through the ridge of a timber pitched roof covered with double lap tiling and show how the margin is calculated. [*Fig 48*]

21 Two general methods of covering hips in double lap tiling are in used. Describe or sketch these two methods showing clearly their difference. [*Fig 49*]

22 By means of a fully dimensioned and annotated sketch show a typical verge detail using double lap tiling. [*Fig 50*]

23 Briefly describe single lap tiling and by means of a neat sketch show a typical single lap tile. [*Fig 51*]

24 To a suitable scale draw a fully dimensioned and annotated section through the eaves of a timber pitched roof using single lap tiling. [*Fig 52*]

25 Briefly describe roof slates and slating and sketch a fully annotated detail through the ridge of a timber pitched roof covered with head nailed slates. [*Fig 53*]

26 In the context of thermal insulation to roofs the roof can be constructed as a cold or warm roof. By means of a fully annotated diagram show one of these arrangements. [*Fig 54*]

27 Flat roofs are usually constructed to a fall of below 10°. Show by comparative diagrams three different methods of obtaining the necessary fall.
[*Fig 55*]

28 To a suitable scale draw a fully dimensioned and annotated diagram through the eaves of a typical flat timber roof with a three layer built-up roofing felt finish.
[*Fig 56*]

29 By means of a fully annotated detail show a typical section through the verge of a small span timber flat roof with three layer built-up roofing felt finish. [*Fig 57*]

30 By means of a fully annotated detail show a typical section at an abutment of a small span timber flat roof and a brick wall where the roof covering is two coat asphalt. [*Fig 58*]

31 By means of neat sketches and by reference to Part K of the Building Regulations show what is meant by a zone of open space in the context of windows to habitable rooms. [*Fig 59*]

32 Window schedules can be presented in a tabulated format. On a sheet of A4 paper design a typical window schedule which could be reproduced and used for any situation and make some typical entries on the schedule to show its use. [*Fig 60*]

33 Describe in detail three types of glass which could be specified for domestic work.
[*Fig 61*]

34 By means of comparative details show the difference between glazing with putty and glazing with beads. [*Fig 62*]

1 A corner header brick is called a:
(a) coin. (b) quoin; (c) queen closer; (d) half bat. [*Fig 29*]

2 The number of facing bricks per m² required for English Bond is:
(a) 89, (b) 98; (c) 58; (d) 79. [*Fig 30*]

3 The number of facing bricks per m² required for Flemish Bond is:
(a) 65; (b) 79; (c) 97; (d) 98. [*Fig 31*]

4 In a half brick wall in stretcher bond the maximum spacing between attached piers is:
(a) 1500 mm; (b) 2000 mm; (c) 2500 mm; (d) 3000 mm. [*Fig 33*]

5 The overhang of a coping stone to a parapet wall should be not less than:
(a) 25 mm; (b) 32 mm; (c) 40 mm; (d) 50 mm. [*Fig 34*]

6 A damp-proof course over a lintel should extend beyond the lintel by at least:
(a) 100 mm; (b) 150 mm; (c) 200 mm; (d) 300 mm. [*Fig 35*]

7 Wedge bricks used in arch construction are called:
(a) voussoirs; (b) intrados; (c) extrados; (d) wedge closers. [*Fig 36*]

8 The centre brick in an arch is called:
(a) abutment; (b) voussoir; (c) key brick; (d) springer. [*Fig 36*]

9 Brick arches are constructed over timber profiles called centres and these should be:
(a) same width as wall;
(b) wider than wall thickness;
(c) less than wall thickness;
(d) project on one side of wall and flush with the other side. [*Fig 37*]

10 Openings in walls consist of three components namely head, sill and:
(a) jambs; (b) lintel; (c) base; (d) muntin. [*Fig 38*]

11 If slates are used as a vertical damp-proof course in an opening in a cavity wall the number of courses should be:
(a) 1; (b) 2; (c) 3, (d) slates should not be used. [*Fig 39*]

12 The minimum recommended overhang of an external sill to an opening is:
(a) 38 mm; (b) 25 mm; (c) 83 mm; (d) 65 mm. [*Fig 40*]

13 A jack rafter spans between:
(a) ridge and eaves;   (c) hip and eaves;
(b) ridge and verge;   (d) another name for a purlin. [*Fig 41*]

14 In timber pitched roofs the traditional size for the wall plate is:
(a) 32 X 200 mm;   (c) 75 X 75 mm;
(b) 100 X 50 mm.   (d) 100 X 75 mm; [*Fig 42*]

15 In timber pitched roofs the rafters should be fixed below the top of the ridge a minimum distance of:
(a) 10 mm; (b) 15 mm; (c) 25 mm; (d) 32 mm. [*Fig 43*]

16 Toothed plate timber connectors are used in conjunction with:
(a) bolts; (b) glue; (c) screws; (d) copper nails. [*Fig 44*]

17 In trussed rafters the steel plate connectors are used:
(a) on one face of the joint;
(b) to both faces of the joint;
(c) between abutting surfaces of the joint;
(d) not suitable for trussed rafters. [*Fig 45*]

18 When fixing a roof underlay the minimum head lap should be:
(a) 50 mm; (b) 75 mm; (c) 100 mm; (d) 150 mm. [*Fig 46*]

19 The length of a standard plain tile is:
(a) 300 mm; (b) 200 mm; (c) 165 mm; (d) 265 mm. [*Fig 47*]

20 In setting out double lap tiling the margin is equal to the:
(a) lap; (b) gauge; (c) length of tile; (d) width of tile. [*Fig 48*]

21 Hip tiles should be bedded in a cement mortar mix of:
(a) 1:3; (b) 1:1; (c) 3:1; (d) 1:1:6. [*Fig 49*]

22 The length of a soaker for double lap tiling should be:
(a) 100 mm; (b) 75 mm; (c) gauge + lap; (d) gauge – lap. [*Fig 50*]

23 In single lap tiling the lap should be:
(a) 100 mm;   (c) tile length – tile width;
(b) 75 mm;   (d) tile length – margin. [*Fig 51*]

24 The minimum recommended pitch for roof slating is:
(a) 10°; (b) 20°; (c) 25°, (d) 35°. [*Fig 53*]

25 In the context of thermal insulation to roofs a warm roof has:
(a) insulation at rafter level;   (c) central heating;
(b) insulation at ceiling level;   (d) no thermal insulation at all. [*Fig 54*]

26 In the context of flat roof coverings the usual minimum pitch for sheet coverings is:
(a) 1:30; (b) 1:60; (c) 1:80; (d) 1:100. [*Fig 55*]

27 In timber flat roofs strutting should be used between the joists if the span exceeds:
(a) 1200 mm; (b) 2000 mm; (c) 2400 mm; (d) 4200 mm. [*Fig 56*]

28 At an abutment the skirting to a flat roof should have a minimum height of:
(a) 100 mm; (b) 150 mm; (c) 50 mm; (d) 400 mm. [*Fig 57*]

29 The minimum recommended distance for a side lap in built-up roofing felt is:
(a) 50 mm; (b) 75 mm; (c) 100 mm; (d) 150 mm [*Fig 58*]

30 For natural ventilation of a habitable room some part of the ventilator should be a distance above the floor level of at least:
(a) 1200 mm; (b) 3600 mm; (c) 1150 mm; (d) 1750 mm. [*Fig 59*]

31 The minimum recommended depth of rebate for general glazing is:
(a) 2 mm; (b) 3 mm; (c) 6 mm; (d) 10 mm. [*Fig 61*]

32 When deciding on the width of a rebate for glass secured by beads the width should be:
(a) bead + glass + 3 mm: (c) bead + glass + 10 mm;
(b) bead + glass + 6 mm;   (d) bead + glass. [*Fig 62*]

Bonding ~ an arrangement of bricks in a wall, column or pier laid to a set pattern to maintain an adequate lap.

Purposes of Brick Bonding ~

1. Obtain maximum strength whilst distributing the loads to be carried throughout the wall, column or pier.
2. Ensure lateral stability and resistance to side thrusts.
3. Create an acceptable appearance.

Lap Forms ~

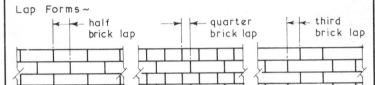

HALF BONDING
used in half brick
thick walls built in
stretcher bond

QUARTER BONDING
used in most bonds
built with standard

THIRD BONDING
used in bonds built
with metric bricks

Simple Bonding Rules ~

1. Bond is set out along length of wall working from each end to ensure that no vertical joints are above one another in consecutive courses.

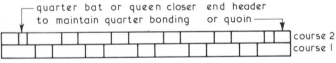

NB all odd numbered courses set out as course 1 and all even numbered courses set out as course 2

2. Walls which are not in exact bond length can be set out thus –

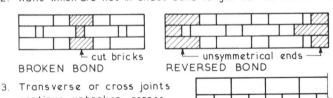

BROKEN BOND

REVERSED BOND

3. Transverse or cross joints continue unbroken across the width of wall unless stopped by a face stretcher.

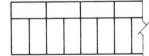

Fig 29   Brick bonding — principles

English Bond ~ formed by laying alternate courses of stretchers and headers it is one of the strongest bonds but it will require more facing bricks than other bonds (89 facing bricks per m²)

Typical Example ~

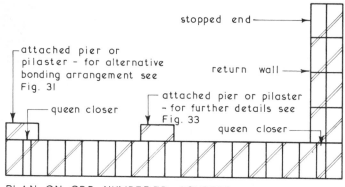

stopped end

attached pier or pilaster - for alternative bonding arrangement see Fig. 31

return wall

attached pier or pilaster – for further details see Fig. 33

queen closer

queen closer

PLAN ON ODD NUMBERED COURSES

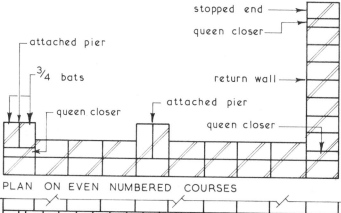

stopped end

queen closer

attached pier

³⁄₄ bats

return wall

queen closer

attached pier

queen closer

PLAN ON EVEN NUMBERED COURSES

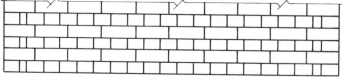

ELEVATION

Fig 30   Brick bonding — English bond

43

Flemish Bond ~ formed by laying headers and stretchers alternately in each course. Not as strong as English bond but is considered to be aesthetically superior uses less facing bricks. (79 facing brick per m²)

Typical Example ~

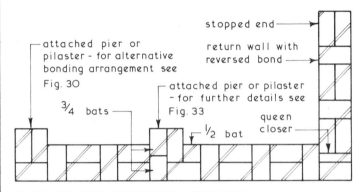

PLAN ON ODD NUMBERED COURSES

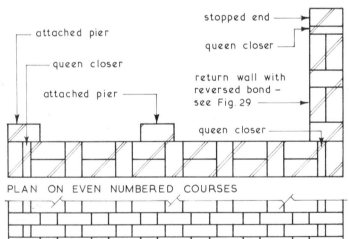

PLAN ON EVEN NUMBERED COURSES

ELEVATION

Fig 31 Brick bonding — Flemish bond

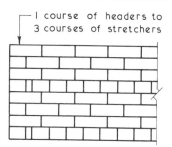

I course of headers to 3 courses of stretchers

ENGLISH GARDEN WALL
BOND - gives quick lateral
spread of load - uses less
facings than English bond.

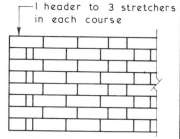

I header to 3 stretchers
in each course

FLEMISH GARDEN WALL
BOND - enables a fair face
to be kept on both sides
of a one brick thick wall.

ENGLISH CROSS BOND - header placed next to end
stretcher in every other stretcher course which thus
staggers stretchers enabling patterns or diapers to be
picked out in different texture or coloured bricks.

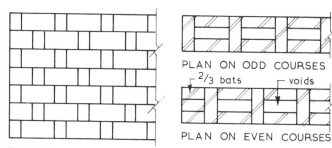

PLAN ON ODD COURSES

2/3 bats   voids

PLAN ON EVEN COURSES

RAT TRAP BOND - uses brick on edge courses - hollow
pockets or voids reduce total weight of wall and by
the bricks on edge there is an overall saving of materials.

Fig 32   Brick bonding — special bonds

Attached Piers ~ the main function of an attached pier is to give lateral support to the wall of which it forms part from the base to the top of the wall. It also has the subsidiary function of dividing a wall into distinct lengths whereby each length can be considered as a wall. Generally walls must be tied at end to an attached pier, buttressing or return wall.

Typical Examples ~

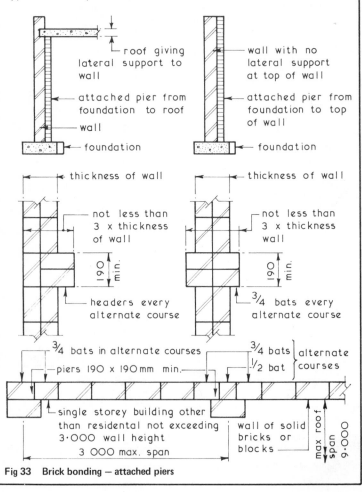

- roof giving lateral support to wall
- attached pier from foundation to roof
- wall
- foundation

- wall with no lateral support at top of wall
- attached pier from foundation to top of wall
- foundation

thickness of wall

- not less than 3 x thickness of wall

190 min.

- headers every alternate course

thickness of wall

- not less than 3 x thickness wall

190 min.

- ¾ bats every alternate course

- ¾ bats in alternate courses
- piers 190 x 190 mm min.
- ¾ bats ⎱ alternate
- ½ bat ⎰ courses

- single storey building other than residental not exceeding 3·000 wall height
3 000 max. span

wall of solid bricks or blocks

max roof span 9·000

Fig 33   Brick bonding — attached piers

Parapet ~ a low wall projecting above the level of a roof, bridge or balcony forming a guard or barrier at the edge. Parapets are exposed to the elements on three faces namely front, rear and top and will therefore need careful design and construction if they are to be durable and reliable.

Typical Details ~

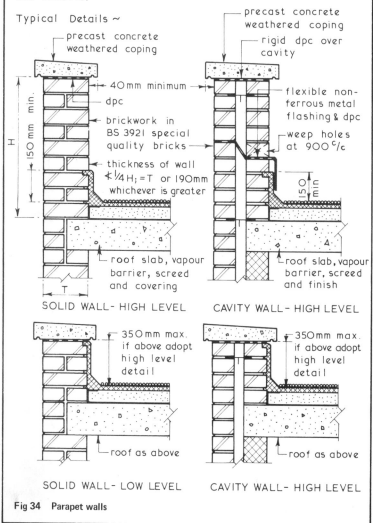

- precast concrete weathered coping
- 40 mm minimum
- dpc
- brickwork in BS 3921 special quality bricks
- thickness of wall $\not< \frac{1}{4}$H; = T or 190 mm whichever is greater
- roof slab, vapour barrier, screed and covering

150 mm min.

H

T

SOLID WALL- HIGH LEVEL

- precast concrete weathered coping
- rigid dpc over cavity
- flexible non-ferrous metal flashing & dpc
- weep holes at 900 c/c
- roof slab, vapour barrier, screed and finish

150 min

CAVITY WALL- HIGH LEVEL

- 350 mm max. if above adopt high level detail
- roof as above

SOLID WALL- LOW LEVEL

- 350mm max. if above adopt high level detail
- roof as above

CAVITY WALL- HIGH LEVEL

Fig 34   Parapet walls

47

Function ~ the primary function of any damp-proof course (dpc) or damp-proof membrane (dpm) is to provide an impermeable barrier to the passage of moisture. The three basic ways in which damp-proof courses are used is to ~

1. Resist moisture penetration from below (rising damp.)
2. Resist moisture penetration from above.
3. Resist moisture penetration from horizontal entry.

Typical Examples ~

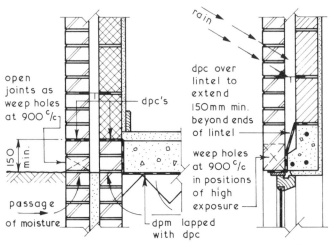

open joints as weep holes at 900 c/c

dpc's

150 min.

passage of moisture

dpm lapped with dpc

PENETRATION FROM BELOW
(Ground Floor / Externa Wall)

rain

dpc over lintel to extend 150mm min. beyond ends of lintel

weep holes at 900 c/c in positions of high exposure

PENETRATION FROM ABOVE
(Window / Door Head)

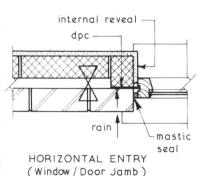

internal reveal

dpc

rain

mastic seal

HORIZONTAL ENTRY
(Window / Door Jamb)

SUITABLE DPC MATERIALS

Engineering bricks - BS 3921
Slates − BS 3798
Lead − BS 1178
Copper − BS 2870
Bitumen based products
Propriety emulsions
Polythene
Pitch polymers
Mastic asphalt − BS 1079 & 1418
BS 743 − Materials for dpc's

Fig 35   Damp-proof courses and membranes

Supports Over Openings ~ the primary function of any support over an opening is to carry the loads above the opening and transmit them safely to the abutments, jambs or piers on both sides. A support over an opening is usually required since the opening infilling such as a door or window frame will not have sufficient strength to carry the load through its own members.

Types of Support ~

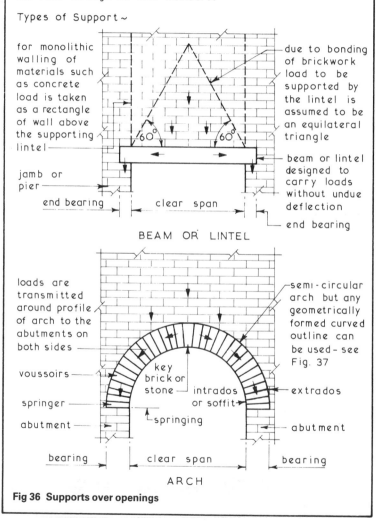

for monolithic walling of materials such as concrete load is taken as a rectangle of wall above the supporting lintel

jamb or pier

end bearing

clear span

due to bonding of brickwork load to be supported by the lintel is assumed to be an equilateral triangle

beam or lintel designed to carry loads without undue deflection

end bearing

BEAM OR LINTEL

loads are transmitted around profile of arch to the abutments on both sides

voussoirs

springer

abutment

bearing

clear span

bearing

semi-circular arch but any geometrically formed curved outline can be used – see Fig. 37

key brick or stone

intrados or soffit

springing

extrados

abutment

ARCH

Fig 36 Supports over openings

49

Arch Construction ~ by the arrangement of the bricks or stones in an arch over an opening it will be self supporting once the jointing material has set and gained adequate strength. The arch must therefore be constructed over a temporary support until the arch becomes self supporting. The traditional method is to use a framed timber support called a centre. Permanent arch centres are also available for small spans and simple formats.

Typical Arch Formats ~

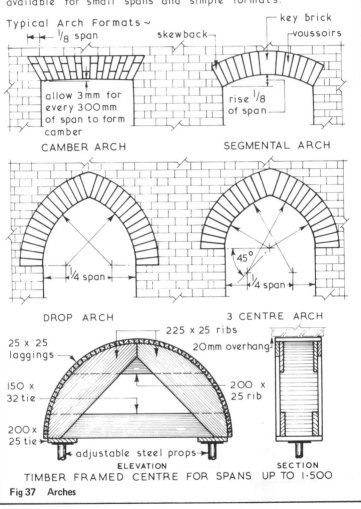

CAMBER ARCH

SEGMENTAL ARCH

DROP ARCH

3 CENTRE ARCH

TIMBER FRAMED CENTRE FOR SPANS UP TO 1·500

Fig 37   Arches

Openings ~ these consist of a head, jambs and sill and the different methods and treatments which can be used in their formation is very wide but they are all based on the same concepts.

Typical Head Details ~

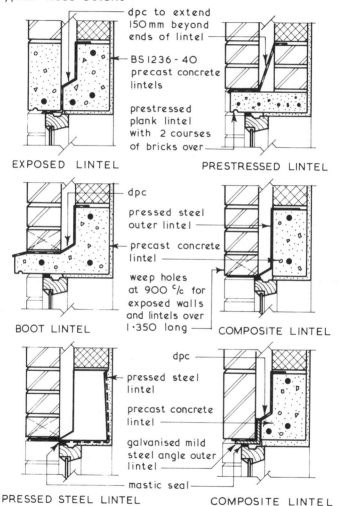

dpc to extend 150 mm beyond ends of lintel

BS 1236 - 40 precast concrete lintels

prestressed plank lintel with 2 courses of bricks over

EXPOSED LINTEL

PRESTRESSED LINTEL

dpc

pressed steel outer lintel

precast concrete lintel

weep holes at 900 c/c for exposed walls and lintels over 1·350 long

BOOT LINTEL

COMPOSITE LINTEL

pressed steel lintel

precast concrete lintel

galvanised mild steel angle outer lintel

mastic seal

dpc

PRESSED STEEL LINTEL

COMPOSITE LINTEL

Fig 38   Opening details 1 — heads

Jambs ~ these may be bonded as in solid walls or unbonded as in cavity walls. The latter must have some means of preventing the ingress of moisture from the outer leaf to the inner leaf and hence the interior of the building.

Typical Jamb Details ~

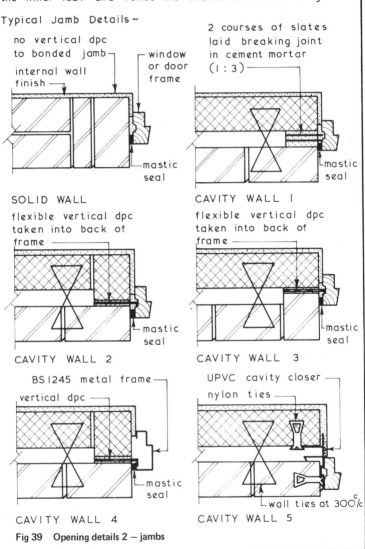

no vertical dpc to bonded jamb

internal wall finish

window or door frame

mastic seal

SOLID WALL

2 courses of slates laid breaking joint in cement mortar (1:3)

mastic seal

CAVITY WALL 1

flexible vertical dpc taken into back of frame

mastic seal

CAVITY WALL 2

flexible vertical dpc taken into back of frame

mastic seal

CAVITY WALL 3

BS 1245 metal frame

vertical dpc

mastic seal

CAVITY WALL 4

UPVC cavity closer

nylon ties

wall ties at 300 c/c

CAVITY WALL 5

Fig 39   Opening details 2 — jambs

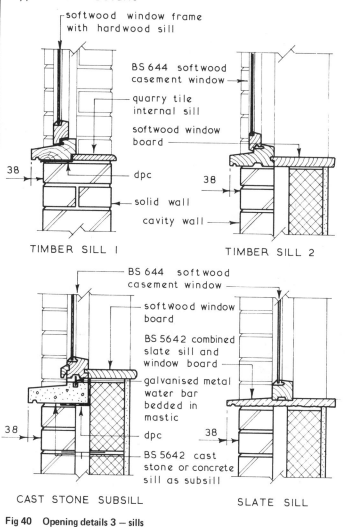

Sills ~ the primary function of any sill is to collect the rainwater which has run down the face of the window or door and shed it clear of the wall below.

Typical Sill Details ~

softwood window frame with hardwood sill

BS 644 softwood casement window

quarry tile internal sill

softwood window board

dpc

solid wall

cavity wall

TIMBER SILL 1

38

TIMBER SILL 2

38

BS 644 softwood casement window

softwood window board

BS 5642 combined slate sill and window board

galvanised metal water bar bedded in mastic

dpc

BS 5642 cast stone or concrete sill as subsill

38

CAST STONE SUBSILL

38

SLATE SILL

Fig 40   Opening details 3 — sills

Pitched Roofs ~ the primary functions of any domestic roof are to -
1. Provide an adequate barrier to the penetration of the elements.
2. Maintain the internal environment by providing an adequate resistance to heat loss.

A roof is in a very exposed situation and must therefore be designed and constructed in such a manner as to -
1. Safely resist all imposed loadings such as snow and wind.
2. Be capable of accommodating thermal and moisture movements.
3. Be durable so as to give a satisfactory performance and reduce maintenance to a minimum.

Component Parts of a Pitched Roof ~

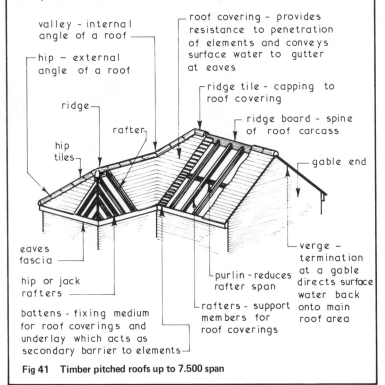

valley - internal angle of a roof

hip - external angle of a roof

roof covering - provides resistance to penetration of elements and conveys surface water to gutter at eaves

ridge

rafter

ridge tile - capping to roof covering

ridge board - spine of roof carcass

hip tiles

gable end

eaves fascia

hip or jack rafters

purlin - reduces rafter span

rafters - support members for roof coverings

verge - termination at a gable directs surface water back onto main roof area

battens - fixing medium for roof coverings and underlay which acts as secondary barrier to elements

Fig 41   Timber pitched roofs up to 7.500 span

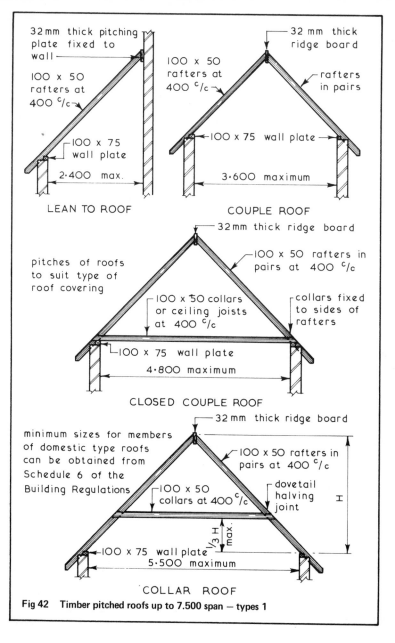

Fig 42   Timber pitched roofs up to 7.500 span — types 1

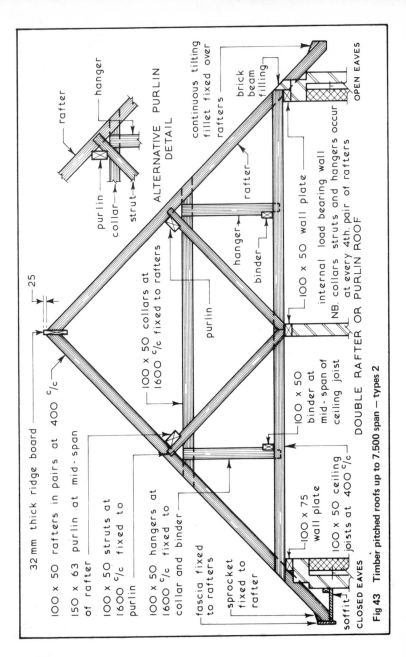

Fig 43 Timber pitched roofs up to 7.500 span — types 2

Labels on figure:

- 32 mm thick ridge board
- 100 × 50 rafters in pairs at 400 c/c
- 150 × 63 purlin at mid-span of rafter
- 100 × 50 struts at 1600 c/c fixed to purlin
- 100 × 50 hangers at 1600 c/c fixed to collar and binder
- 25
- rafter
- hanger
- purlin
- collar
- strut
- ALTERNATIVE PURLIN DETAIL
- continuous tilting fillet fixed over rafters
- brick beam filling
- OPEN EAVES
- hanger
- rafter
- binder
- 100 × 50 wall plate
- internal load bearing wall
- NB. collars struts and hangers occur at every 4th. pair of rafters
- DOUBLE RAFTER OR PURLIN ROOF
- 100 × 50 collars at 1600 c/c fixed to rafters
- purlin
- 100 × 50 binder at mid-span of ceiling joist
- 100 × 75 wall plate
- 100 × 50 ceiling joists at 400 c/c
- fascia fixed to rafters
- sprocket fixed to rafter
- soffit
- CLOSED EAVES

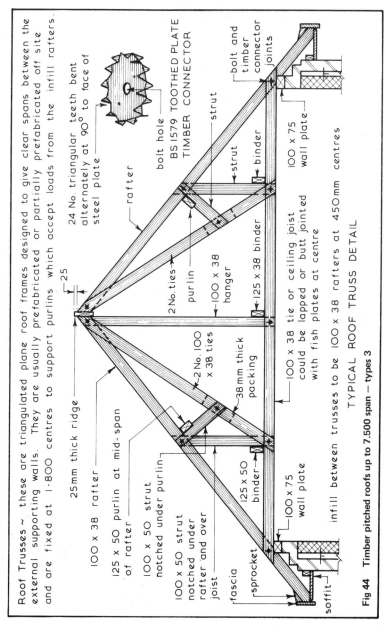

Roof Trusses ~ these are triangulated plane roof frames designed to give clear spans between the external supporting walls. They are usually prefabricated or partially prefabricated off site and are fixed at 1·800 centres to support purlins which accept loads from the infill rafters.

25 mm thick ridge

100 x 38 rafter

125 x 50 purlin at mid-span of rafter

100 x 50 strut notched under purlin

100 x 50 strut notched under rafter and over joist

fascia

sprocket

soffit

100 x 75 wall plate

125 x 50 binder

2 No. 100 x 38 ties

38 mm thick packing

rafter

2 No. ties

purlin

100 x 38 hanger

125 x 38 binder

24 No. triangular teeth bent alternately at 90° to face of steel plate

bolt hole

BS 1579 TOOTHED PLATE TIMBER CONNECTOR

strut

strut

binder

100 x 75 wall plate

bolt and timber connector joints

100 x 38 tie or ceiling joist could be lapped or butt jointed with fish plates at centre

infill between trusses to be 100 x 38 rafters at 450mm centres

TYPICAL ROOF TRUSS DETAIL

Fig 44   Timber pitched roofs up to 7.500 span — types 3

57

Trussed Rafters ~ these are triangulated plane roof frames designed to give clear spans between the external supporting walls. They are delivered to the site as a prefabricated component where they are fixed to the wall plate at 600mm centres. Trussed rafters do not require any ridge board or purlins since they receive their lateral stability by using larger tiling battens than those used on traditional roofs.

galvanised steel truss plate connectors formed by cutting, punching and bending a series of spikes at right angles to face. Plates are inserted under heavy pressure to both faces of trussed rafter at all butt joints

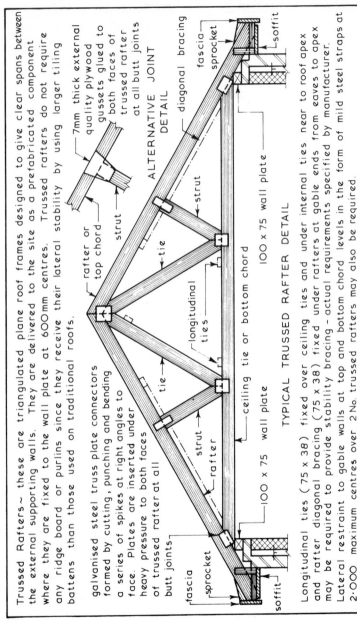

7mm thick external quality plywood gussets glued to both faces of trussed rafter at all butt joints

rafter or top chord

strut

ALTERNATIVE JOINT DETAIL

diagonal bracing

fascia

sprocket

soffit

strut

tie

longitudinal ties

100 x 75 wall plate

ceiling tie or bottom chord

tie

strut

rafter

100 x 75 wall plate

fascia

sprocket

soffit

TYPICAL TRUSSED RAFTER DETAIL

Longitudinal ties (75 x 38) fixed over ceiling ties and under internal ties near to roof apex and rafter diagonal bracing (75 x 38) fixed under rafters at gable ends from eaves to apex may be required to provide stability bracing - actual requirements specified by manufacturer. Lateral restraint to gable walls at top and bottom chord levels in the form of mild steel straps at 2·000 maximum centres over 2 No. trussed rafters may also be required.

Fig 45 Timber pitched roofs up to 7.500 span ~ types 4

Roof Underlays ~ sometimes called sarking or roofing felt provides the barrier to the entry of wind, snow and rain blown between the tiles or slates it also prevents the entry of water from capillary action.

Suitable Materials ~

Bitumen fibre based felts
Bitumen asbestos based felts
Bitumen glass fibre based felts

supplied in rolls 1m wide × 10 or 20m long to the recommendations of BS 747

Sheathing and Hair felts − supplied in rolls 810 mm wide × 25 m long to the recommendations of BS 747

Plastic Sheeting underlays – these are lighter, require less storage space, have greater flexibility at low temperatures and high resistance to tearing but have a greater risk to the formation of condensation than the BS 747 felts and should not be used on roof pitches below 20.°

Typical Details ~

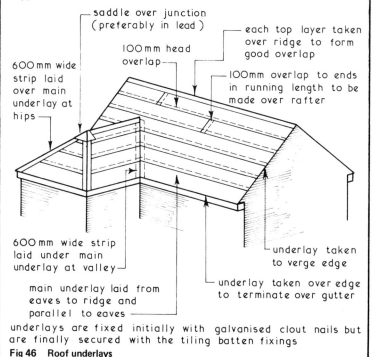

- saddle over junction (preferably in lead)

100 mm head overlap

- each top layer taken over ridge to form good overlap

600 mm wide strip laid over main underlay at hips

- 100mm overlap to ends in running length to be made over rafter

600 mm wide strip laid under main underlay at valley

main underlay laid from eaves to ridge and parallel to eaves

- underlay taken to verge edge

- underlay taken over edge to terminate over gutter

underlays are fixed initially with galvanised clout nails but are finally secured with the tiling batten fixings

Fig 46   Roof underlays

Double Lap Tiles~ these are the traditional tile covering for pitched roofs and are available made from clay and concrete and are usually called plain tiles. Plain tiles have a slight camber in their length to ensure that the tail of the tile will bed and not ride on the tile below. There is always at least two layers of tiles covering any part of the roof. Each tile has at least two nibs on the underside of its head so that it can be hung on support battens nailed over the rafters. Two nail holes provide the means of fixing the tile to the batten, in practice only every 4th. course of tiles is nailed unless the roof exposure is high. Double lap tiles are laid to a bond so that the edge joints between the tiles are in the centre of the tiles immediately below and above the course under consideration.

Typical Plain Tile Details ~

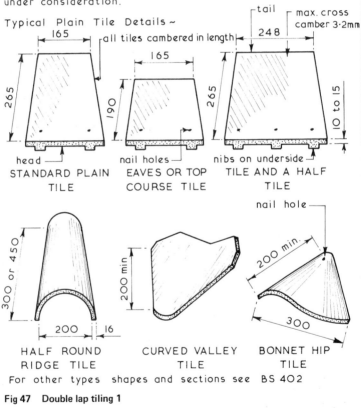

STANDARD PLAIN TILE

EAVES OR TOP COURSE TILE

TILE AND A HALF TILE

HALF ROUND RIDGE TILE

CURVED VALLEY TILE

BONNET HIP TILE

For other types shapes and sections see BS 402

Fig 47 Double lap tiling 1

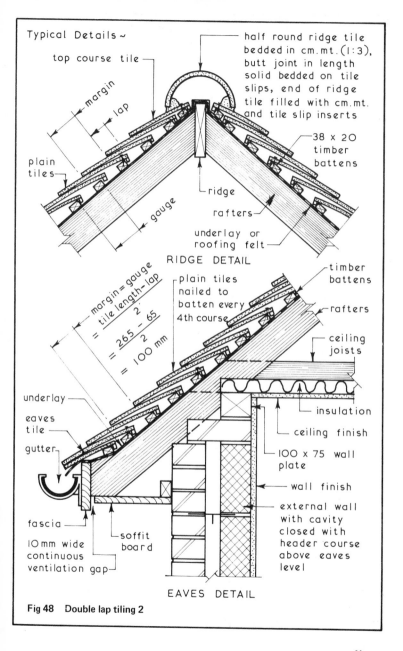

Typical Details ~

top course tile

half round ridge tile bedded in cm.mt. (1:3), butt joint in length solid bedded on tile slips, end of ridge tile filled with cm.mt. and tile slip inserts

margin

lap

plain tiles

38 x 20 timber battens

ridge

gauge

rafters

underlay or roofing felt

RIDGE DETAIL

margin = gauge = $\dfrac{\text{tile length} - \text{lap}}{2}$

= $\dfrac{265 - 65}{2}$

= 100 mm

plain tiles nailed to batten every 4th course

timber battens

rafters

ceiling joists

underlay

eaves tile

gutter

fascia

10 mm wide continuous ventilation gap

soffit board

insulation

ceiling finish

100 x 75 wall plate

wall finish

external wall with cavity closed with header course above eaves level

EAVES DETAIL

Fig 48   Double lap tiling 2

61

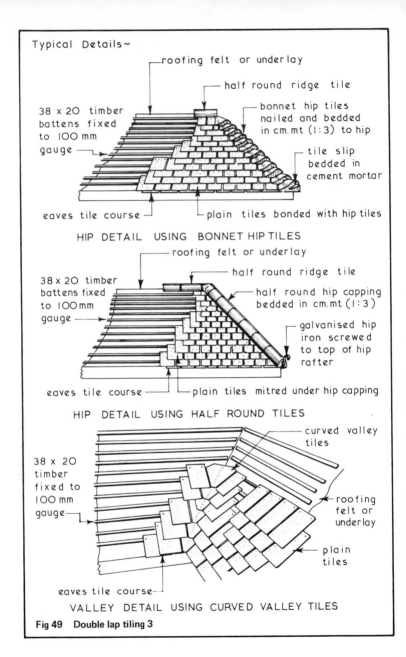

Typical Details~

roofing felt or underlay

half round ridge tile

38 x 20 timber battens fixed to 100 mm gauge

bonnet hip tiles nailed and bedded in cm.mt (1:3) to hip

tile slip bedded in cement mortar

eaves tile course

plain tiles bonded with hip tiles

HIP DETAIL USING BONNET HIP TILES

roofing felt or underlay

half round ridge tile

38 x 20 timber battens fixed to 100mm gauge

half round hip capping bedded in cm.mt (1:3)

galvanised hip iron screwed to top of hip rafter

eaves tile course

plain tiles mitred under hip capping

HIP DETAIL USING HALF ROUND TILES

curved valley tiles

38 x 20 timber fixed to 100 mm gauge

roofing felt or underlay

plain tiles

eaves tile course

VALLEY DETAIL USING CURVED VALLEY TILES

Fig 49   Double lap tiling 3

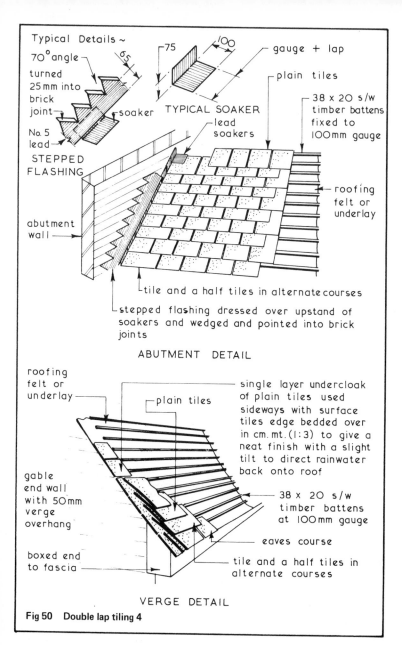

Typical Details ~

70° angle turned 25 mm into brick joint

65

No. 5 lead

STEPPED FLASHING

soaker

75 · 100 · gauge + lap

TYPICAL SOAKER

plain tiles

38 x 20 s/w timber battens fixed to 100mm gauge

lead soakers

abutment wall

roofing felt or underlay

tile and a half tiles in alternate courses

stepped flashing dressed over upstand of soakers and wedged and pointed into brick joints

ABUTMENT DETAIL

roofing felt or underlay

plain tiles

single layer undercloak of plain tiles used sideways with surface tiles edge bedded over in cm. mt. (1:3) to give a neat finish with a slight tilt to direct rainwater back onto roof

gable end wall with 50mm verge overhang

38 x 20 s/w timber battens at 100mm gauge

eaves course

boxed end to fascia

tile and a half tiles in alternate courses

VERGE DETAIL

Fig 50   Double lap tiling 4

63

Single Lap Tiling ~ so called because the single lap of one tile over another provides the weather tightness as opposed to the two layers of tiles used in double lap tiling. Most of the single lap tiles produced in clay and concrete have a tongue and groove joint along their side edges and in some patterns on all four edges which forms a series of interlocking joints and therefore these tiles are called single lap interlocking tiles. Generally there will be an overall reduction in the weight of the roof covering when compared with double lap tiling but the batten size is larger than that used for plain tiles and as a minimum every tile in alternate courses should be twice nailed although a good specification will require every tile to be twice nailed. The gauge or batten spacing for single lap tiling is found by subtracting the end lap from the length of the tile.

Typical Single Lap Tiles ~

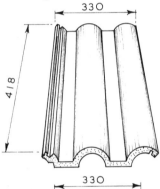

ROLL TYPE TILE

minimum pitch 30°

head lap 75mm

side lap 30mm

gauge 343mm

linear coverage 300mm

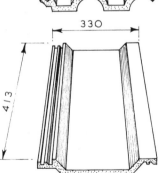

TROUGH TYPE TILE

minimum pitch 15°

head lap 75mm

side lap 38mm

gauge 338mm

linear coverage 292mm

Fig 51   Single lap tiling   1

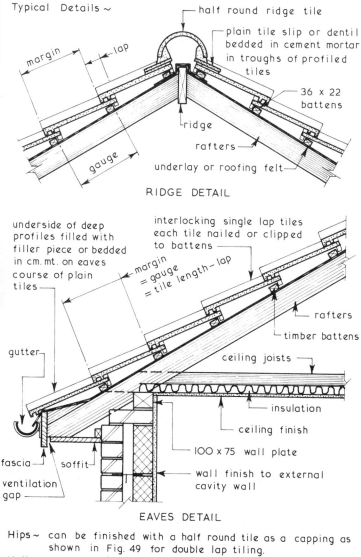

Typical Details ~

half round ridge tile

plain tile slip or dentil
bedded in cement mortar
in troughs of profiled
tiles

margin

lap

36 x 22
battens

ridge

gauge

rafters

underlay or roofing felt

RIDGE DETAIL

underside of deep
profiles filled with
filler piece or bedded
in cm. mt. on eaves
course of plain
tiles

interlocking single lap tiles
each tile nailed or clipped
to battens

margin
= gauge = tile length – lap

rafters

timber battens

gutter

ceiling joists

insulation

ceiling finish

100 x 75 wall plate

fascia

soffit

wall finish to external
cavity wall

ventilation
gap

EAVES DETAIL

Hips ~ can be finished with a half round tile as a capping as
shown in Fig. 49 for double lap tiling.

Valleys ~ these can be finished by using special valley trough
tiles or with a lead lined gutter - see manufacturer's data.

Fig 52 Single lap tiling 2

Slates ~ slate is a natural dense material which can be split into thin sheets and cut to form a small unit covering suitable for pitched roofs in excess of 25° pitch. Slates are graded according to thickness and texture the thinnest being known as 'Bests'. Slates are laid to the same double lap principles as plain tiles. Ridges and hips are normally covered with half round or angular tiles whereas valley junctions are usually of mitred slates over soakers. Unlike plain tiles every slate in every course is fixed to the battens by head nailing or centre nailing, the latter being used on long slates and on pitches below 35° to overcome the problem of vibration caused by the wind which can break head nailed long slates. For full range of sizes and gradings see BS 680.

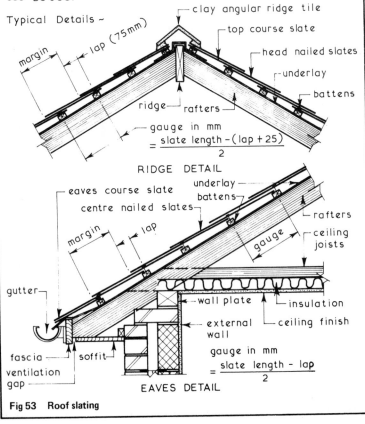

Typical Details ~

gauge in mm
$$= \frac{\text{slate length} - (\text{lap} + 25)}{2}$$

RIDGE DETAIL

gauge in mm
$$= \frac{\text{slate length} - \text{lap}}{2}$$

EAVES DETAIL

Fig 53   Roof slating

66

Thermal Insulation ~ this is required in most roofs to reduce the heat loss from the interior of the building which will create a better internal environment reducing the risk of condensation and give a saving on heating costs. In domestic situations Part F of the Building Regulations gives a maximum allowable thermal transmission or U value of 0·6 W/m² °C. This can be achieved by placing thermal insulating material(s) at rafter level or at ceiling level, the former creating a warm roof void and the latter a cold roof void. A combination of both methods is also possible.

Typical Details ~

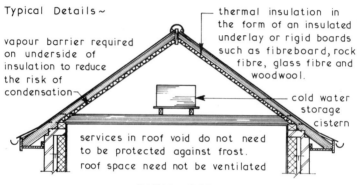

thermal insulation in the form of an insulated underlay or rigid boards such as fibreboard, rock fibre, glass fibre and woodwool.

vapour barrier required on underside of insulation to reduce the risk of condensation

cold water storage cistern

services in roof void do not need to be protected against frost. roof space need not be ventilated

WARM ROOF

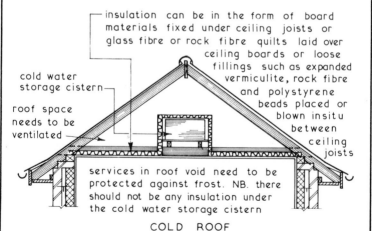

insulation can be in the form of board materials fixed under ceiling joists or glass fibre or rock fibre quilts laid over ceiling boards or loose fillings such as expanded vermiculite, rock fibre and polystyrene beads placed or blown insitu between ceiling joists

cold water storage cistern

roof space needs to be ventilated

services in roof void need to be protected against frost. NB. there should not be any insulation under the cold water storage cistern

COLD ROOF

Fig 54   Thermal insulation to roofs

Flat Roofs ~ these roofs are very seldom flat with a pitch of 0° but are considered to be flat if the pitch does not exceed 10.° The actual pitch chosen can be governed by the roof covering selected and /or by the required rate of rainwater discharge off the roof. As a general rule the minimum pitch for smooth surfaces such as asphalt should be 1:80 or 0°-43' and for sheet coverings with laps 1:60 or 0°-57.'

Methods of Obtaining Falls ~

1. Joists cut to falls

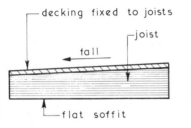

Simple to fix but could be wasteful in terms of timber unless two joists are cut from one piece of timber

2. Joists laid to falls

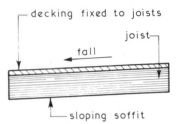

Economic and simple but sloping soffit may not be acceptable but this could be hidden by a flat suspended ceiling

3. Firrings with joist run

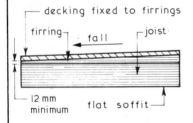

Simple and effective but does not provide a means of natural cross ventilation. Usual method employed.

4. Firrings against joist run

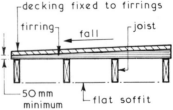

Simple and effective but uses more timber than 3 but does provide a means of natural cross ventilation

Wherever possible joists should span the shortest distance of the roof plan

Fig 55 Timber flat roofs up to 4.0000 span—1

Timber Roof Joists ~ the spacing and sizes of joists is related to the loadings and span actual dimensions for domestic loadings can be taken direct from Schedule 6 of the Building Regulations or they can be calculated from first principles in the same manner as used for timber upper floors. Strutting between joists should be used if the span exceeds 2·400 to restrict joist movements and twisting.

Typical Eaves Details ~

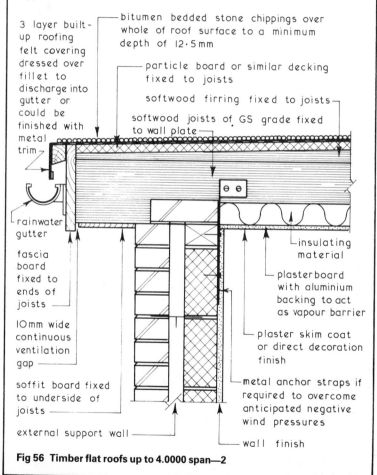

3 layer built-up roofing felt covering dressed over fillet to discharge into gutter or could be finished with metal trim

bitumen bedded stone chippings over whole of roof surface to a minimum depth of 12·5mm

particle board or similar decking fixed to joists

softwood firring fixed to joists

softwood joists of GS grade fixed to wall plate

rainwater gutter

fascia board fixed to ends of joists

10mm wide continuous ventilation gap

soffit board fixed to underside of joists

external support wall

insulating material

plasterboard with aluminium backing to act as vapour barrier

plaster skim coat or direct decoration finish

metal anchor straps if required to overcome anticipated negative wind pressures

wall finish

Fig 56 Timber flat roofs up to 4.0000 span—2

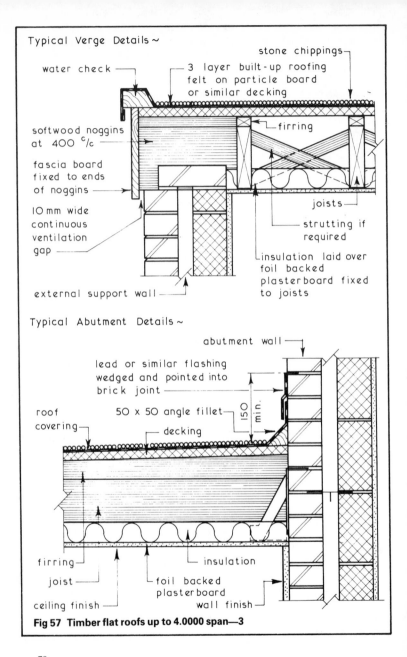

Typical Verge Details ~

stone chippings

water check

3 layer built-up roofing felt on particle board or similar decking

softwood noggins at 400 c/c

firring

fascia board fixed to ends of noggins

joists

10 mm wide continuous ventilation gap

strutting if required

external support wall

insulation laid over foil backed plasterboard fixed to joists

Typical Abutment Details ~

abutment wall

lead or similar flashing wedged and pointed into brick joint

roof covering

50 x 50 angle fillet

decking

150 min.

firring

insulation

joist

foil backed plasterboard

ceiling finish

wall finish

Fig 57 Timber flat roofs up to 4.0000 span—3

Built-up Roofing Felt ~ this consists of three layers of bitumen roofing felt to BS 747 and should be laid to the recommendations of CP 144 Part 3. The layers of felt are bonded together with hot bitumen and should have staggered laps of 50mm minimum for side laps and 75mm minimum for end laps - for typical details see Figs. 56 & 57. Other felt materials which could be used are the two layer polyester based roofing felts which use a non-woven polyester base instead of the woven base used in the BS 747 felts.

Mastic Asphalt ~ this consists of two layers of mastic asphalt laid breaking the joints and built up to a minimum thickness of 20mm and should be laid to the recommendations of CP 144 Part 4. The mastic asphalt is laid over an isolating membrane of black sheathing felt complying with BS 747A(i) which should be laid loose with 50mm minimum overlaps.

Typical Details ~

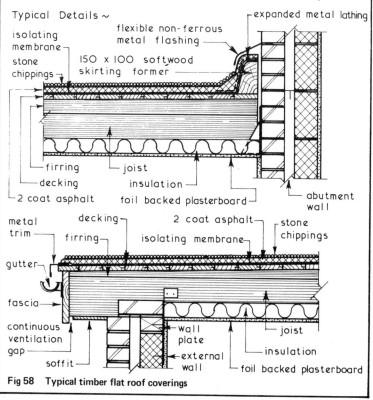

Fig 58 Typical timber flat roof coverings

71

Functions ~ the main functions of any window are to -

1. Provide natural daylight to the interior of the building.
2. Provide a means of natural ventilation.
3. Provide a barrier to the elements.
4. Provide a vision out giving the occupants a visual contact with the world outside.

If windows are fitted into the walls of a habitable room they must comply with the minimum requirements set out in Part K of the Building Regulations which requires that :-

1. Total area of ventilator(s) is greater than $^1/20$ of the floor area of the room or rooms it serves.
2. Some part of the ventilator is at least 1·750 above the floor level.
3. A zone of open space is maintained in front of the window thus :-

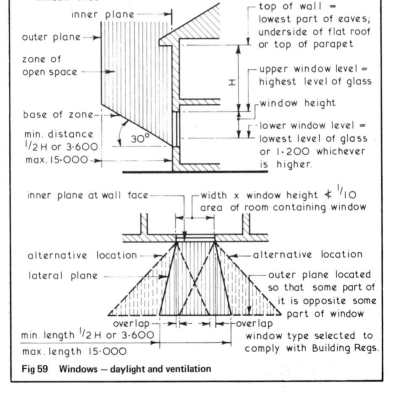

Fig 59   Windows — daylight and ventilation

Schedules ~ the main function of a schedule is to collect together all the necessary information for a particular group of components such as windows, doors and drainage inspection chambers. There is no standard format for schedules but they should be easy to read, accurate and contain all the necessary information for their purpose. Schedules are usually presented in a tabulated format which can be related to and read in conjunction with the working drawings.

Typical Example ~

WINDOW SCHEDULE – Sheet 1 of 1				Drawn By: RC		Date: 14/4/81	Rev.	
Contract Title & Number: Lane End Farm - H 341/80						Drg. Nos. C (31) 450 - 7		
Location or Number	Type	Material	Overall Size	Glass	Ironmongery		Sill	
							External	Internal
1 & 2	9FCV4 - BS 990 Subframe - BS 1285	steel softwood	910 x 1214 970 x 1275	146 x 1140 632 x 553 670 x 594 3mm float	supplied with casements		2 cos. plain tiles	150 x 150 x 15 quarry tiles
3, 4, 5 & 6	240V - BS 644 Part 1	softwood	1206 x 1206	480 x 280 580 x 700 480 x 1030 3mm float	ditto		ditto	25 mm thick softwood
7	Purpose made - Drg. No.C(31)-457	softwood	1770 x 1600	460 x 200 1080 x 300 460 x 1040 1080 x 1140 3mm float	1 - 200 mm 1 - 300 mm al. stays 1 - al. alloy fastener		sill of frame	ditto

Fig 60 Windows — schedules

Glass ~ this material is produced by fusing together soda, lime and silica with other minor ingredients such as magnesia and alumina. A number of glass types are available for domestic work and these include :–

Clear Float ~ used where clear undistorted vision is required. Available thicknesses range from 3mm to 25mm.

Clear Sheet ~ suitable for all clear glass areas but because the two faces of the glass are never perfectly flat or parallel some distortion of vision usually occurs. This type of glass is gradually being superseded by the clear float glass. Available thicknesses range from 3mm to 6mm.

Translucent Glass ~ these are patterned glasses most having one patterned surface and one relatively flat surface. The amount of obscurity and diffusion obtained depend on the type and nature of pattern. Available thicknesses range from 4mm to 6mm for patterned glasses and from 5mm to 10mm for rough cast glasses.

Wired Glass ~ obtainable as a clear polished wired glass or as a rough cast wired glass with a nominal thickness of 7mm. Generally used where a degree of fire resistance is required. Georgian wired glass has a 12mm square mesh whereas the hexagonally wired glass has a 20mm mesh.

Choice of Glass ~ the main factors to be considered are :–
1. Resistance to wind loadings. 2. Clear vision required.
3. Privacy. 4. Security. 5. Fire resistance. 6. Aesthetics.

Glazing Terminology ~

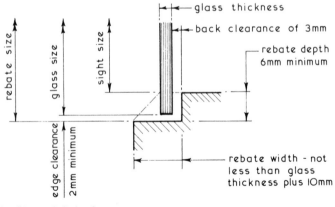

Fig 61 Glass and glazing 1

Glazing ~ the act of fixing glass into a frame or surround. In domestic work this is usually achieved by locating the glass in a rebate and securing it with putty or beading and should be carried out in accordance with the recommendations contained in CP 152.

Timber Surrounds ~ linseed oil putty to BS 544 – rebate to be clean, dry and primed before glazing is carried out. Putty should be protected with paint within two weeks of application.

Metal Surrounds ~ metal casement putty if metal surround is to be painted – if surround is not to be painted a non-setting compound should be used.

Typical Glazing Details ~

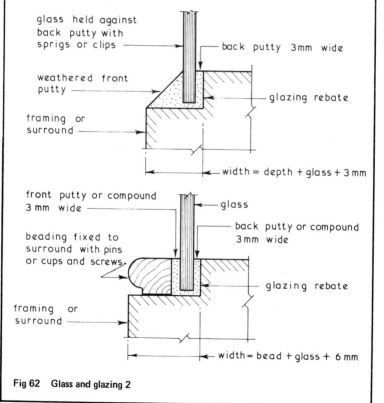

glass held against back putty with sprigs or clips

back putty 3 mm wide

weathered front putty

glazing rebate

framing or surround

width = depth + glass + 3 mm

front putty or compound 3 mm wide

glass

back putty or compound 3 mm wide

beading fixed to surround with pins or cups and screws

glazing rebate

framing or surround

width = bead + glass + 6 mm

Fig 62   Glass and glazing 2

To check answers to the following questions the student should refer to the information given in the figure number(s) quoted at the end of each question.

SHORT ANSWER QUESTIONS

1 Define the term 'simply supported slab.'   [*Fig 63*]

2 A simply supported reinforced concrete slab with a rectangular plan shape usually has two layers of reinforcing bars. Name the two types of bars used and state which type is placed nearest to the outer fibres of the slab.   [*Fig 64*]

3 List the stages of construction for a simply supported reinforced concrete slab.
                                                                      [*Fig 65*]

4 What is the main purpose of a concrete floor screed?   [*Fig 66*]

5 Floating concrete floor screeds are laid on a resilient quilt. What is the main objective of this form of screed?   [*Fig 67*]

6 Define the term 'timber stud partition.'   [*Fig 68*]

7 In timber stairs intermediate landings can be introduced to enable a change of direction to be made. Name and define two such landing types.   [*Fig 69*]

8 In the context of timber stairs define the terms, 'rebated nosing' and 'pendant newel.'   [*Fig 70*]

9 What are the main functions of a door?   [*Fig 71*]

10 When using plasterboard for ceilings two sizes can be considered. Name and define these two types of plasterboard.   [*Fig 72*]

11 Spray plasters are really a means of application rather than a type of plaster. Comment on this statement.   [*Fig 73*]

12 Glazed wall tiles come in many formats. By means of neat sketches show two such formats.   [*Fig 74*]

13 In the context of internal wall tiling define and describe grouting.   [*Fig 75*]

14 State the main functions of a paint.   [*Fig 76*]

15 State what is involved in preparing a surface to receive paint and why this activity is so important.   [*Fig 77*]

1 By means of comparative diagrams show the condition of a simply supported concrete slab before and after loading, assuming that the induced tensile stresses caused by the loading will exceed the tensile resistance of the concrete slab.
[*Fig 63*]

2 To a suitable scale draw a fully dimensioned and annotated section through a simply supported reinforced concrete slab 150 mm thick having a clear span of 4.500 in both directions and a 300 mm overhang of the one brick thick supporting wall. The reinforcement pattern must be clearly shown. [*Fig 64*]

3 Describe or illustrate the complete construction sequence of a simply supported reinforced concrete slab. [*Fig 65*]

4 Describe and illustrate as necessary two different types of concrete floor screed.
[*Figs 66 and 67*]

5 A timber stud partition with a central door opening is required to partition off a room having a width of 3.000 and a ceiling height of 2.400. To a suitable scale draw a fully dimensioned and annotated elevation of a suitable partition clearly showing the method of construction. [*Fig 68*]

6 To a suitable scale draw the plans of a typical dog-leg stair and an open newel stair. [*Fig 69*]

7 To a suitable scale draw a fully dimensioned and annotated section through a typical intermediate timber stair landing. [*Fig 70*]

8 State the main factors to be considered when selecting an internal door. By means of a neat sketch show the detail of a typical door lining for a lightweight internal door. [*Fig 71*]

9 By means of comparative details show the difference in construction of a plaster-board ceiling at the wall abutment with and without the use of a cove finish.
[*Fig 72*]

10 Most plasters produced in this country comply with BS 1191 which lists four classes. Name and describe three of these classes. [*Fig 74*]

11 Briefly describe the manufacture of internal glazed wall tiles and show by neat sketches what is meant by bead fittings. [*Fig 74*]

12 Wall tiles can be fixed by thin or thick bedding techniques. Describe in detail both of these methods. [*Fig 75*]

13 Describe the basic composition of paints used in building works. [*Fig 76*]

14 Describe the basic build-up of paint coats and give details of two methods of paint application. [*Fig 77*]

1 The maximum allowed deflection of a simply supported reinforced concrete slab after completion is:
(a) 1/240; (b) 1/300; (c) 1/360; (d) 1/400. [*Fig 63*]

2 In a simply supported reinforced concrete slab the minimum cover of concrete over the main reinforcement should be:
(a) 10 mm; (b) 15 mm; (c) 25 mm; (d) 150 mm. [*Fig 64*]

3 The cover to the reinforcement in a concrete slab is maintained during construction by:
(a) spacers;   (c) plinths of aggregate;
(b) wire clips,   (d) distribution bars. [*Fig 65*]

4 A monolithic concrete floor screed is:
(a) precast;   (c) laid on concrete within 3 hours of
(b) laid on a cured floor;   casting;
                                (d) fixed to timber battens. [*Fig 66*]

5 Unbonded concrete floor screeds are usually specified to have a minimum thickness of:
(a) 40 mm; (b) 50 mm; (c) 65 mm; (d) 100 mm. [*Fig 67*]

6 In a timber stud partition the horizontal intermediate members are called:
(a) studs; (b) folding wedges; (c) rails, (d) noggins. [*Fig 68*]

7 A half space landing enables a turn in a stair to be made through:
(a) 45°; (b) 90°, (c) 180°; (d) 270°. [*Fig 69*]

8 In timber stairs a half pendant newel is a:
(a) bottom newel;   (c) post fixed to a wall;
(b) storey height newel;   (d) post fixed to the ground floor. [*Fig 70*]

9 The width of a door lining should be:
(a) not less than 150 mm;   (c) twice door thickness + width of door stop;
(b) wall thickness + wall finishes;   (d) width of wall. [*Fig 71*]

10 Plasterboard should be fixed to the ceiling joists with:
(a) special nails 25 mm long;   (c) special glue;
(b) special nails 32 mm long;   (d) special steel screws. [*Fig 72*]

11 The site storage of plaster requires:
(a) dry conditions;   (c) special silo;
(b) damp conditions,   (d) no special provisions. [*Fig 73*]

12 In the context of internal glazed wall tiling the joint between the tiles can be maintained by using spacer tiles which have:
(a) square edges;   (c) edge lugs;
(b) rounded edges;   (d) rebated edges. [*Fig 74*]

13 Water based paints are usually called;
(a) solvents;   (c) emulsion paints;
(b) gloss paints;   (d) non-drip paints. [*Fig 76*]

14 Paint is usually supplied in metal containers ranging in capacity from:
(a) 100 millilitres to 250 millilitres; (c) 250 millilitres to 5 litres;
(b) 250 millilitres to 1 litre;   (d) 5 litres to 250 litres. [*Fig 77*]

Simply Supported Slabs ~ these are slabs which rest on a bearing and for design purposes are not considered to be fixed to the support and are therefore, in theory, free to lift. In practice however they are restrained from unacceptable lifting by their own self weight plus any loadings.

Concrete Slabs ~ concrete is a material which is strong in compression and weak in tension and if the member is overloaded its tensile resistance may be exceeded leading to structural failure.

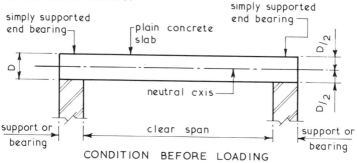

simply supported end bearing

plain concrete slab

simply supported end bearing

D

neutral axis

D/2

D/2

support or bearing

clear span

support or bearing

CONDITION BEFORE LOADING

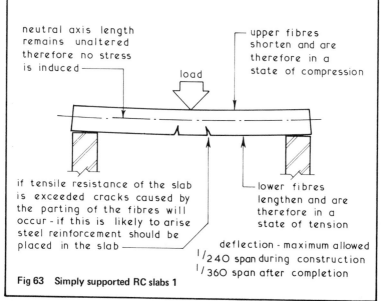

neutral axis length remains unaltered therefore no stress is induced

upper fibres shorten and are therefore in a state of compression

load

if tensile resistance of the slab is exceeded cracks caused by the parting of the fibres will occur - if this is likely to arise steel reinforcement should be placed in the slab

lower fibres lengthen and are therefore in a state of tension

deflection - maximum allowed $1/240$ span during construction $1/360$ span after completion

Fig 63   Simply supported RC slabs 1

Reinforcement~ generally in the form of steel bars which are used to provide the tensile strength which plain concrete lacks. The number, diameter, spacing, shape and type of bars to be used have to be designed, the process of which is beyond the scope of this text. Reinforcement is placed as near to the outside fibres as practicable, a cover of concrete over the reinforcement is required to protect the steel bars from corrosion and to provide a degree of fire resistance. Slabs which are square in plan are considered to be spanning in two directions and therefore main reinforcing bars are used both ways whereas slabs which are rectangular in plan are considered to span across the shortest distance and main bars are used in this direction only with smaller diameter distribution bars placed at right angles forming a mat or grid.

Typical Details~

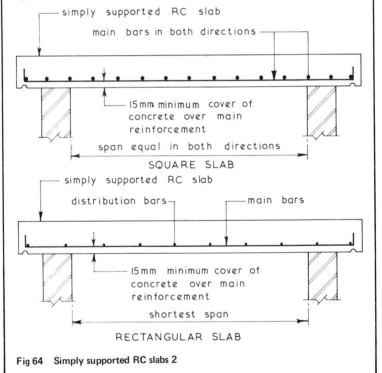

simply supported RC slab

main bars in both directions

15mm minimum cover of concrete over main reinforcement

span equal in both directions

SQUARE SLAB

simply supported RC slab

distribution bars

main bars

15mm minimum cover of concrete over main reinforcement

shortest span

RECTANGULAR SLAB

Fig 64   Simply supported RC slabs 2

Construction~ whatever method of construction is used the construction sequence will follow the same pattern -

1. Assemble and erect formwork.
2. Prepare and place reinforcement.
3. Pour and compact or vibrate concrete.
4. Strike and remove formwork in stages as curing proceeds.

Typical Example ~

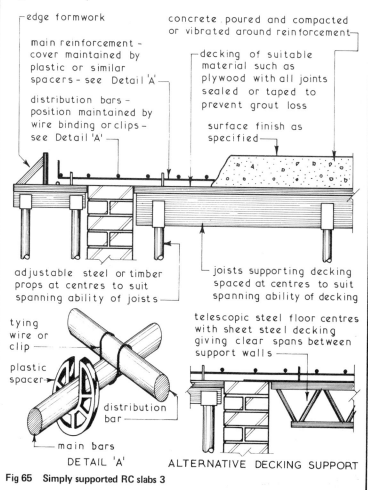

edge formwork

main reinforcement - cover maintained by plastic or similar spacers - see Detail 'A'

distribution bars - position maintained by wire binding or clips - see Detail 'A'

concrete poured and compacted or vibrated around reinforcement

decking of suitable material such as plywood with all joints sealed or taped to prevent grout loss

surface finish as specified

adjustable steel or timber props at centres to suit spanning ability of joists

joists supporting decking spaced at centres to suit spanning ability of decking

tying wire or clip

plastic spacer

distribution bar

main bars

DETAIL 'A'

telescopic steel floor centres with sheet steel decking giving clear spans between support walls

ALTERNATIVE DECKING SUPPORT

Fig 65 Simply supported RC slabs 3

81

Concrete Floor Screeds~ these are used to give a concrete floor a finish suitable to receive the floor finish or covering specified. It should be noted that it is not always necessary or desirable to apply a floor screed to receive a floor covering, techniques are available to enable the concrete floor surface to be prepared at the time of casting to receive the coverings at a later stage.

Typical Screed Mixes~

Screed Thickness	Cement	Dry Fine Agg < 5mm	Coarse Agg. >5mm<10mm
up to 40 mm	1	3 to $4\frac{1}{2}$	—
40 to 75mm	1	3 to $4\frac{1}{2}$	—
	1	$\frac{1}{2}$	3

Laying Floor Screeds~ floor screeds should not be laid in bays since this can cause curling at the edges, screeds can however be laid in 3·000 wide strips to receive thin coverings. Levelling of screeds is achieved by working to levelled timber screeding batten or alternatively a 75mm wide band of levelled screed with square edges can be laid to the perimeter of the floor prior to the general screed laying operation.

Screed Types~

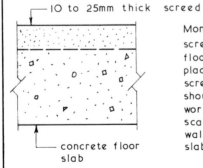

10 to 25mm thick screed

concrete floor slab

Monolithic Screeds —
screed laid directly on concrete floor slab within three hours of placing concrete - before any screed is placed all surface water should be removed – all screeding work should be carried out from scaffold board runways to avoid walking on the 'green' concrete slab.

Fig 66   Concrete floor screeds 1

Screed Types ~

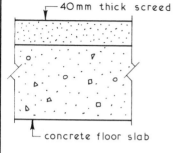

40 mm thick screed

concrete floor slab

Separate Screeds —

screed is laid onto the concrete floor slab after it has cured. The floor surface must be clean and rough enough to ensure an adequate bond unless the floor surface is prepared by applying a suitable bonding agent or by brushing with a cement / water grout of a thick cream like consistency just before laying the screed.

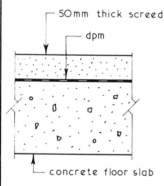

50 mm thick screed

dpm

concrete floor slab

Unbonded Screeds —

screed laid directly over a damp-proof membrane – care must be taken during this operation to ensure that the damp-proof membrane is not damaged.

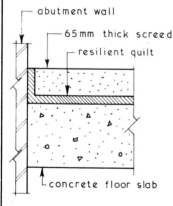

abutment wall

65 mm thick screed

resilient quilt

concrete floor slab

Floating Screeds —

a resilient quilt of 25 mm thickness is laid with butt joints and turned up at the edges against the abutment walls the screed being laid directly over the quilt. The main objective of this form of floor screed is to improve the sound insulation properties of the floor.

Fig 67   Concrete floor screeds 2

Timber Stud Partitions ~ these are non-load bearing internal dividing walls which are easy to construct, lightweight, adaptable and can be clad and infilled with various materials to give different finishes and properties. The timber studs should be of prepared or planed material to ensure that the wall is of constant thickness with parallel faces. Stud spacings will be governed by the size and spanning ability of the facing or cladding material.

Typical Details ~

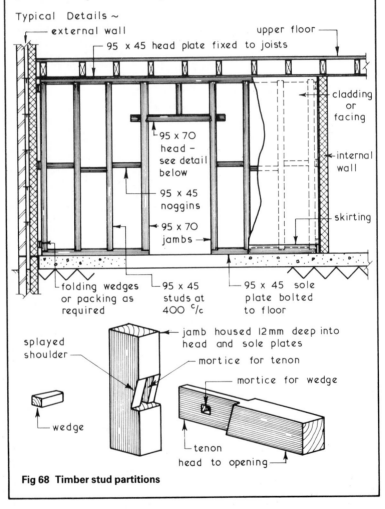

- external wall
- upper floor
- 95 x 45 head plate fixed to joists
- cladding or facing
- 95 x 70 head - see detail below
- internal wall
- 95 x 45 noggins
- skirting
- 95 x 70 jambs
- folding wedges or packing as required
- 95 x 45 studs at 400 c/c
- 95 x 45 sole plate bolted to floor

jamb housed 12mm deep into head and sole plates

splayed shoulder

mortice for tenon

mortice for wedge

wedge

tenon

head to opening

Fig 68 Timber stud partitions

Timber Stairs ~ these must comply with the minimum requirements set out in Part H of the Building Regulations. Straight flight stairs are simple, easy to construct and install but by the introduction of intermediate landings stairs can be designed to change direction of travel and be more compact in plan than the straight flight stairs.

Landings ~ these are designed and constructed in the same manner as timber upper floors but due to the shorter spans they require smaller joist sections. Landings can be detailed for a 90° change of direction (quarter space landing) or a 180° change of direction (half space landing) and can be introduced at any position between the two floors being served by the stairs.

Typical Layouts ~

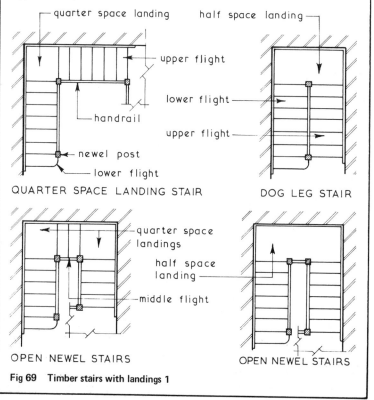

QUARTER SPACE LANDING STAIR

DOG LEG STAIR

OPEN NEWEL STAIRS

OPEN NEWEL STAIRS

Fig 69   Timber stairs with landings 1

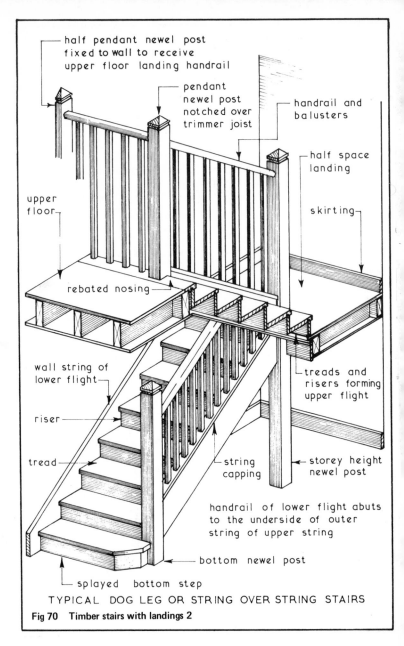

half pendant newel post
fixed to wall to receive
upper floor landing handrail

pendant
newel post
notched over
trimmer joist

handrail and
balusters

half space
landing

upper
floor

skirting

rebated nosing

wall string of
lower flight

treads and
risers forming
upper flight

riser

string
capping

tread

storey height
newel post

handrail of lower flight abuts
to the underside of outer
string of upper string

bottom newel post

splayed bottom step

TYPICAL DOG LEG OR STRING OVER STRING STAIRS

Fig 70 Timber stairs with landings 2

86

Functions ~ the main functions of any door are:-

1. Provide a means of access and egress.
2. Maintain continuity of wall function when closed.
3. Provide a degree of privacy and security.

Choice of door type can be determined by :-

1. Position - whether internal or external
2. Properties required - fire resistant, glazed to provide for borrowed light or vision through, etc.
3. Appearance - flush or panelled, painted or polished, etc.

Door Schedules ~ these can be prepared in the same manner and for the same purpose as that given for windows in Fig. 60.

Internal Doors ~ these are usually lightweight and can be fixed to a lining, if heavy doors are specified these can be hung to frames in a similar manner to external doors. An alternative method is to use door sets which are usually storey height and supplied with prehung doors.

Typical Door Lining Details ~

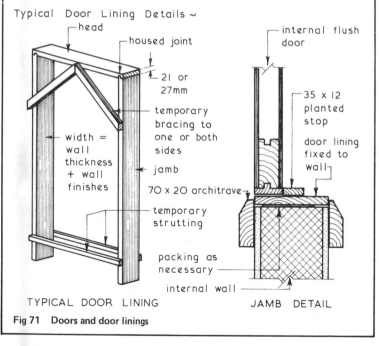

TYPICAL DOOR LINING   JAMB DETAIL

Fig 71 Doors and door linings

87

Plasterboard~ this is a rigid board made with a core of gypsum sandwiched between face sheets of strong durable paper. In the context of ceilings two sizes can be considered –

1. Baseboard – 2·400 x 1·200 x 9·5mm thick for supports at centres not exceeding 400mm; 2·400 x 1·200 x 12·7mm for supports at centres not exceeding 600mm. Baseboard has square edges and therefore the joints will need reinforcing with jute scrim at least 90mm wide or alternatively a special tape to prevent cracking.

2. Gypsum Lath – 1·200 x 406 x 9·5 or 12·7mm thick. Lath has rounded edges which eliminates the need to reinforce the joints.

Both types of board are available with an aluminium foil face which acts as a vapour barrier and/or a reflective surface to give an insulating plasterboard, for both uses the joints must be sealed with a self adhesive aluminium strip.

The boards are fixed to the underside of the floor or ceiling joists with galvanised or sheradised plasterboard nails at not more than 150mm centres and are laid breaking the joint. Edge treatments consist of jute scrim reinforcement or a preformed plaster cove moulding.

Typical Details ~

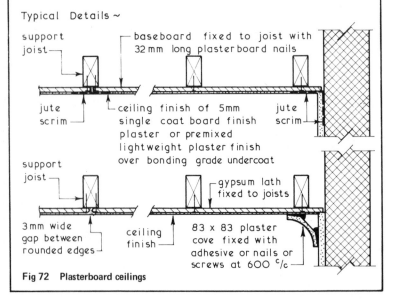

Fig 72   Plasterboard ceilings

Plaster Types~ most plasters produced in this country comply with the recommendations of BS 1191 which lists 4 classes:-

Class A - Plaster of Paris - rapid setting (approximately 10 minutes), can be mixed with a little sand or used neat. Suitable for repairs, mouldings and as a filler.

Class B - Retarded Hemi-hydrate Plaster - largest group of plasters available, they set fairly quickly to give a reasonably hard impact resistant surface. Class B plasters are available as undercoat, wall finish and board finish plasters. Undercoats can be of a sand/plaster mixture (1:1-3 plaster:sand) the actual mix ratio being governed by the background to which the plaster is being applied. Wall plaster mixes can have an addition of lime (3:1 plaster:lime) to increase the workability of the mix unless being applied to a concrete background when a neat plaster mix should be used. Board finish plasters should always be used neat.

Class C - Anhydrous Plaster - finishing plasters with a slower setting time than Class B plasters and are therefore easier to bring to smooth finish. Class C plasters are usually applied neat.

Class D - Keene's Plaster - these are finishing plasters with a slow setting time giving a very hard, smooth and impact resistant surface making them suitable for arrises and reveals. These plasters are always applied neat.

Lightweight Premixed Plasters - these generally conform to the recommendations for Class B type plasters and have lightweight aggregates such as vermiculite and perlite. Generally they are applied in 2 coats, an undercoat of 11mm thick with a 2mm thick finishing coat. Premixed plasters only require the addition of clean water on site to be ready for use.

Spray Plasters - this is really a means of application rather than a type of plaster since all types of plasters can be successfully applied using a mechanical spray if they are retarded in accordance with the manufacturer's instructions. Some Class B plasters containing a suitable retarder for spray application.

All plasters, like cement, require dry on site storage.

Fig 73   Plasters

Glazed Wall Tiles ~ internal glazed wall tiles are usually made to the recommendations of BS 1281. External glazed wall tiles made from clay or clay/ceramic mixtures are manufactured but there is no British Standard available.

Internal Glazed Wall Tiles ~ the body of the tile can be made from ball-clay, china clay, china stone, flint and limestone. The material is usually mixed with water to the desired consistency, shaped and then fired in a tunnel oven at a high temperature ($1150\,^{\circ}$C) for several days to form the unglazed biscuit tile. The glaze, pattern and colour can now be imparted onto to the biscuit tile before the final firing process at a temperature slightly lower than that of the first firing ($1050\,^{\circ}$C) for about two days.

Typical Internal Glazed Wall Tiles and Fittings ~

Sizes – Modular – 100 x 100 x 5mm thick and 200 x 100 x 6·5mm thick.

Non-modular – 152 x 152 x 5 to 8mm thick and 108 x 108 x 4 and 6·5mm thick.

Fittings – wide range available particularly in the non-modular format.

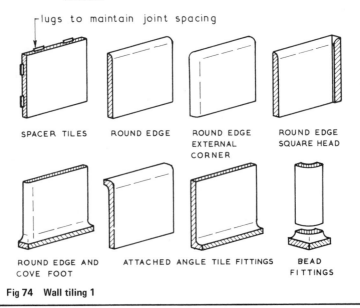

lugs to maintain joint spacing

SPACER TILES   ROUND EDGE   ROUND EDGE EXTERNAL CORNER   ROUND EDGE SQUARE HEAD

ROUND EDGE AND COVE FOOT   ATTACHED ANGLE TILE FITTINGS   BEAD FITTINGS

Fig 74   Wall tiling 1

Bedding of Internal Wall Tiles ~ generally glazed internal wall tiles are considered to be inert in the context of moisture and thermal movement, therefore if movement of the applied wall tile finish is to be avoided attention must be given to the background and the method of fixing the tiles.

Backgrounds ~ these are usually of a cement rendered or plastered surface and should be flat, dry, stable, firmly attached to the substrate and sufficiently old enough for any initial shrinkage to have taken place. The flatness of the background should be not more than 3mm in 2·000 for the thin bedding of tiles and not more than 6mm in 2·000 for thick bedded tiles.

Fixing Wall Tiles ~ two methods are in general use:-

1. Thin Bedding - lightweight internal glazed wall tiles fixed dry using a recommended adhesive which is applied to wall in small areas (1 m²) at a time with a notched trowel, the tile being pressed or tapped into the adhesive.

2. Thick Bedding - cement mortar within the mix range of 1:3 to 1:4 is used as the adhesive either by buttering the backs of the tiles which are then pressed or tapped into position or by rendering the wall surface to a thickness of approximately 10mm and then applying the lightly buttered tiles (1:2 mix) to the rendered wall surface within two hours. It is usually necessary to soak the wall tiles in water to reduce suction before they are placed in position.

Grouting ~ when the wall tiles have set the joints can be grouted by rubbing into the joints a grout paste either using a sponge or brush. Most grouting materials are based on cement with inert fillers and are used neat.

Typical Example ~

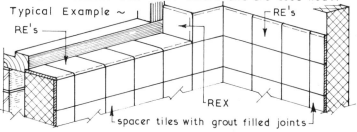

RE's

RE's

REX

spacer tiles with grout filled joints

Fig 75 Wall tiling 2

Functions ~ the main functions of paint are to provide -

1. An economic method of surface protection to building materials and components.
2. An economic method of surface decoration to building materials and components.

Composition ~ the actual composition of any paint can be complex but the basic components are -

1. Binder ~ this is the liquid vehicle or medium which dries to form the surface film and can be composed of linseed oil, drying oils, synthetic resins and water. The first function of a paint medium is to provide a means of spreading the paint over the surface and at the same time acting as a binder to the pigment.

2. Pigment – this provides the body, colour, durability and and corrosion protection properties of the paint. White lead pigments are very durable and moisture resistant but are poisonous and their use is generally restricted to priming and undercoating paints. If a paint contains a lead pigment the fact must be stated on the container. The general pigment used in paint is titanium dioxide which is not poisonous and gives good obliteration of the undercoats.

3. Solvents and Thinners – these are materials which can be added to a paint to alter its viscosity.

Paint Types - there is a wide range available but for most general uses the following can be considered –

1. Oil Based Paints – these are available in priming, undercoat and finishing grades. The latter can be obtained in a wide range of colours and finishes such as matt, semi-matt, eggshell, satin, gloss and enamel. Polyurethane paints have a good hardness and resistance to water and cleaning. Oil based paints are suitable for most applications if used in conjunction with correct primer and undercoat.

2. Water Based Paints – most of these are called emulsion paints the various finishes available being obtained by adding to the water medium additives such as alkyd resin & polyvinyl acetate (PVA). Finishes include matt, eggshell, semi-gloss and gloss. Emulsion paints are easily applied, quick drying and can be obtained with a washable finish and are suitable for most applications.

Fig 76   Paints and painting 1

Supply ~ paint is usually supplied in metal containers ranging from 250 millilitres to 5 litres capacity to the colour ranges recommended in BS 381C (colours for specific purposes) and BS 4800 (paint colours for building purposes).

Application ~ paint can be applied to almost any surface providing the surface preparation and sequence of paint coats are suitable. The manufacturers specification and/or the the recommendations of CP 231 (painting of buildings) should be followed. Preparation of the surface to receive the paint is of the utmost importance since poor preparation is one of the chief causes of paint failure. The preparation consists basically of removing all dirt, grease, dust and ensuring that the surface will provide an adequate key for the paint which is to be applied. In new work the basic build-up of paint coats consists of –

1. Priming Coats – these are used on unpainted surfaces to obtain the necessary adhesion and to inhibit corrosion of ferrous metals. New timber should have the knots treated with a solution of shellac or other alcohol based resin called knotting prior to the application of the primer.

2. Undercoats – these are used on top of the primer after any defects have been made good with a suitable stopper or filler. The primary function of an undercoat is to give the opacity and build-up necessary for the application of the finishing coat(s).

3. Finish – applied directly over the undercoating in one or more coats to impart the required colour and finish.

Paint can applied by :-

1. Brush – the correct type, size and quality of brush such as those recommended in BS 2992 needs to be selected and used. To achieve a first class finish by means of brush application requires a high degree of skill.

2. Spray – as with brush application a high degree of skill is required to achieve a good finish. Generally compressed air sprays or airless sprays are used for building works.

3. Roller – simple and inexpensive method of quickly and cleanly applying a wide range of paints to flat and textured surfaces. Roller heads vary in size from 50 to 450 mm wide with various covers such as sheepskin, synthetic pile fibres, mohair and foamed polystyrene. All paint applicators must be thoroughly cleaned after use.

Fig 77   Paints and painting 2

5 Services and external works

To check answers to the following questions the student should refer to the information given in the figure number(s) quoted at the end of each question.

1 What are the main advantages and disadvantages of a direct cold water system?
[*Fig 78*]

2 Briefly describe an indirect cold water supply system. [*Fig 79*]

3 Name the two valves which should be attached to the boiler in a simple hot water installation. [*Fig 80*]

4 In an indirect hot water system two cisterns are used. Name these two cisterns and state their minimum storage capacities. [*Figs 80 and 81*]

5 In a hot and cold water installation valves are used to control the flow of water. Name three different types of valve and state their respective functions. [*Fig 82*]

6 Define the term 'cold water storage cistern' and state any limitations as to the siting of a cistern. [*Fig 83*]

7 What is an indirect cylinder? [*Fig 84*]

8 What is the basic difference between manipulative and non-manipulative compression joint fittings? [*Fig 85*]

9 Briefly describe fireclay sinks and state how they can be fixed. [*Fig 86*]

10 Briefly describe bath and shower fittings and state how they can be fixed. [*Fig 87*]

11 What is the basic difference between a wash down water closet and a siphonic water closet? [*Fig 88*]

12 Briefly define what is a single stack drainage system. [*Fig 89*]

13 In the context of simple drainage what is an inspection chamber and what materials could be used for its main construction? [*Fig 90*]

14 By means of neat diagrams show how the depth and fall of a trench for drains can be maintained during excavations. [*Fig 91*]

15 State the factors which govern the stability of a highway or road. [*Fig 92*]

16 Footpaths to highways can be constructed in many forms. Name three such forms and state the minimum recommended fall and width for a public footpath.[*Fig 93*]

17 In the context of building works state what could be involved in a landscaping operation.   [*Fig 94*]

18 List the materials which could be involved in the construction of a simple small span bridge.   [*Fig 95*]

LONG ANSWER QUESTIONS

1 By means of an annotated line diagram draw the layout of a typical domestic direct cold water installation from the stop valve on the rising main.   [*Fig 78*]

2 By means of an annotated line diagram draw the layout of a typical domestic indirect cold water installation from the stop valve on the rising main.   [*Fig 79*]

3 By means of an annotated line diagram draw the layout of a typical domestic direct hot water installation from the inlet to the cold water storage cistern.   [*Fig 80*]

4 By means of an annotated line diagram draw the layout of a typical domestic indirect hot water installation from the inlet to the cold water storage cistern.
[*Fig 81*]

5 Ball valves are used to control the flow of water to a storage cistern. Describe or sketch a typical ballvalve stating or showing how it operates.   [*Fig 82*]

6 By means of a clear and fully annotated detail show the section through a typical cold water storage cistern installation.   [*Fig 83*]

7 By means of a neat sketch and description show the working of a primatic cylinder and state its main advantage over an indirect cylinder.   [*Fig 84*]

8 By means of neat sketches show how copper, steel and PVC cold water pipes can be jointed in their running length.   [*Fig 85*]

9 By means of a neat annotated sketch show a section through a typical ceramic wash basin including the fittings to the waste outlet.   [*Fig 86*]

10 By means of a neat annotated sketch show a typical domestic bath installation including all the necessary fittings.   [*Fig 87*]

11 By means of a neat annotated sketch show a section through a typical ceramic washdown or siphonic water closet including its connection to the soil pipe.
[*Fig 88*]

12 In the context of a single stack drainage system show the connection of a bath waste and a water closet branch to the soil and ventilation pipe. List any limitations and assume both fittings are on the same upper floor.   [*Fig 89*]

13 To a suitable scale draw a fully dimensioned and annotated section through a brick or precast concrete inspection chamber for a 100 mm diameter drain with an invert level of 1.000.   [*Fig 90*]

14 By means of neat annotated diagrams show the difference in the bedding of rigid-jointed pipes and flexible-jointed pipes.   [*Fig 91*]

15 By means of neat annotated diagrams show two different methods of constructing highway subsoil drains. [*Fig 92*]

16 By means of neat annotated sections show typical details of a flexible footpath and a footpath constructed with small unit paving blocks. [*Fig 93*]

17 What is meant by a common service trench? State the types of service which could be used in conjunction with such a trench. [*Fig 94*]

18 By means of an annotated sketch show the basic components of a typical small span bridge. [*Fig 95*]

MULTI-CHOICE QUESTIONS

1 In a direct cold water installation the supply to a bath should have a bore size of:
(a) 13 mm; (b) 25 mm; (c) 20 mm; (d) 114 litres. [*Fig 78*]

2 In an indirect cold water installation where the cold water storage cistern feeds a hot water cylinder the minimum capacity of the cistern should be:
(a) 115 litres; (b) 230 litres; (c) 114 litres; (d) 50 litres. [*Fig 79*]

3 The hot water supply pipe from a cylinder should have a horizontal run before the expansion pipe junction of at least:
(a) 230 mm; (b) 300 mm; (c) 450 mm; (d) 114 mm. [*Fig 80*]

4 In a hot water installation the vent or expansion pipe should be terminated:
(a) above the storage cistern;   (c) to the external air;
(b) below the storage cistern;   (d) into rising main. [*Figs 80 and 81*]

5 A bib tap has a:
(a) vertical inlet: (c) top inlet;
(b) horizontal inlet;   (d) front inlet. [*Fig 82*]

6 The draw-off pipes from a cold water storage cistern should be fixed above the bottom of the cistern by at least:
(a) 50 mm; (b) 25 mm; (c) 80 mm; (d) 40 mm. [*Fig 83*]

7 The cold feed pipe to a primatic cylinder is fixed:
(a) near to the top;   (c) in the middle of the side;
(b) in the crown;   (d) near to the bottom. [*Fig 84*]

8 The grooves in the drainer of a sink are called:
(a) flutes; (b) reeds; (c) overflows; (d) rebates. [*Fig 86*]

9 The usual trap size for a bath is:
(a) 20 mm; (b) 32 mm; (c) 38 mm; (d) 100 mm. [*Fig 87*]

10 The water seal in a water closet should be at least:
(a) 405 mm; (b) 380 mm; (c) 100 mm; (d) 50 mm. [*Fig 88*]

11 The water seals in baths, sinks and basins in a single stack drainage system should be at least:
(a) 38 mm; (b) 75 mm; (c) 100 mm; (d) 50 mm. [*Fig 89*]

12 In inspection chambers the fall to the upper surface of the benching should be:
(a) 1:6: (b) 1:16; (c) 1:60; (d) 1:1. [*Fig 90*]

13 When using drainpipes with a spigot-and socket-joint the socket should be placed:
   (a) in direction of flow;   (c) vertically;
   (b) against direction of flow;   (d) there is no set position.   [*Fig 91*]

14 In the context of highway drainage a catch pit is used in conjunction with:
   (a) surface water drains from   (c) road gullies;
       buildings;   (d) subsoil drains.
   (b) surface water sewers;   [*Fig 92*]

15 In rigid paving expansion joints should be placed at maximum centres of:
   (a) 3.000; (b) 9.000; (c) 10.000; (d) 27.000.   [*Fig 93*]

16 In the context of road signs directive signs give:
   (a) warning;   (c) instruction to be obeyed;
   (b) information;   (d) picture of works in progress. [*Fig 94*]

17 In simple small span bridges the support on either side is called:
   (a) embankment; (b) abutment; (c) column; (d) foundation.   [*Fig 95*]

General ~ when planning or designing any water installation the basic physical laws must be considered :-

1. Water is subject to the force of gravity and will find its own level.

2. To overcome friction within the conveying pipes water which is stored prior to distribution will require to be under pressure and this is normally achieved by storing the water at a level above the level of the outlets. The vertical distance between the levels is usually called the head.

3. Water becomes less dense as its temperature is raised therefore warm water will always displace colder water whether in a closed or open circuit.

Direct Cold Water Systems ~ the cold water is supplied to the outlets at mains pressure the only storage requirement is a small capacity cistern to feed the hot water storage tank. These systems are suitable for districts which have high level reservoirs with a good supply and pressure. Main advantage is that drinking water is available from all cold water outlets, disadvantages include lack of reserve in case of supply cut off, risk of back syphonage due to negative mains pressure and a risk of reduced pressure during peak demand periods.

Typical Direct Cold Water System ~

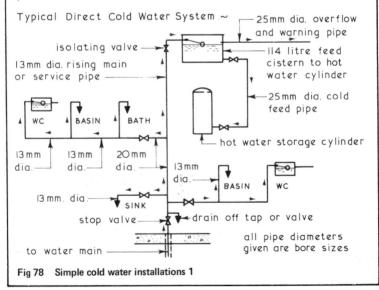

Fig 78   Simple cold water installations 1

Indirect Systems ~ cold water is supplied to all outlets from a cold water storage cistern except for the cold water supply to the sink(s) where the drinking water tap is connected directly to the incoming supply from the main. This system requires more pipework than the indirect system but reduces the risk of back syphonage and provides a reserve of water should the mains supply fail or be cut off. The local water authority will stipulate the system to be used in their area.

Typical Indirect Cold Water System ~

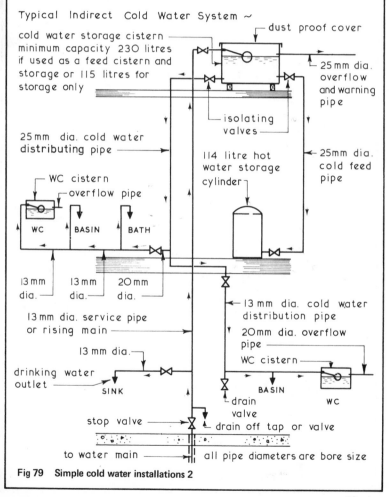

cold water storage cistern — minimum capacity 230 litres if used as a feed cistern and storage or 115 litres for storage only

dust proof cover

25 mm dia. overflow and warning pipe

isolating valves

114 litre hot water storage cylinder

25 mm dia. cold feed pipe

25 mm dia. cold water distributing pipe

WC cistern
overflow pipe

WC   BASIN   BATH

13 mm dia.   13 mm dia.   20 mm dia.

13 mm dia. cold water distribution pipe

20 mm dia. overflow pipe

WC cistern

13 mm dia. service pipe or rising main

13 mm dia.

drinking water outlet

SINK   BASIN   WC

drain valve

stop valve

drain off tap or valve

to water main

all pipe diameters are bore size

Fig 79   Simple cold water installations 2

Direct System ~ this is the simplest and cheapest system of hot water installation. The water is heated in the boiler and the hot water rises by convection to the hot water storage tank or cylinder to be replaced by the cooler water from the bottom of the storage vessel. Hot water drawn from storage is replaced with cold water from the cold water storage cistern. Direct systems are suitable for soft water areas and for installations which are not supplying a central heating circuit.

Typical Direct Hot Water System ~

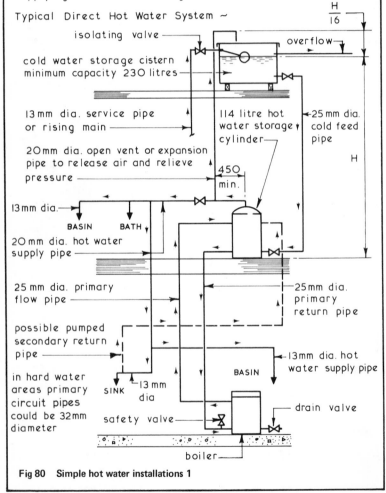

Fig 80   Simple hot water installations 1

Indirect System ~ this is a more complex system than the direct system but does overcome the problem of furring which can occur in direct hot water systems. This method is therefore suitable for hard water areas and in all systems where a central heating circuit is to be part of the hot water installation. Basically the pipe layouts of the two systems are similar but in the indirect system a separate small capacity feed cistern is required to charge and top up the primary circuit. In this system the hot water storage tank or cylinder is in fact a heat exchanger – see Fig. 84.

Typical Indirect Hot Water System ~

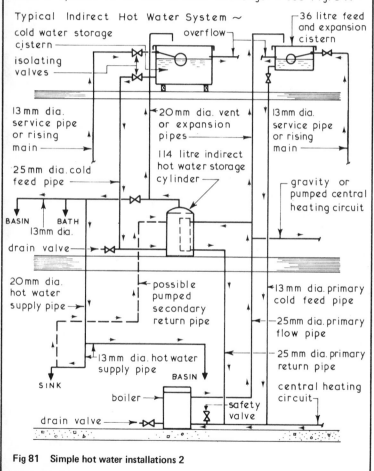

Fig 81   Simple hot water installations 2

Flow Controls ~ these are valves inserted into a water installation to control the water flow along the pipes or to isolate a branch circuit or to control the draw-off of water from the system.

Typical Examples ~

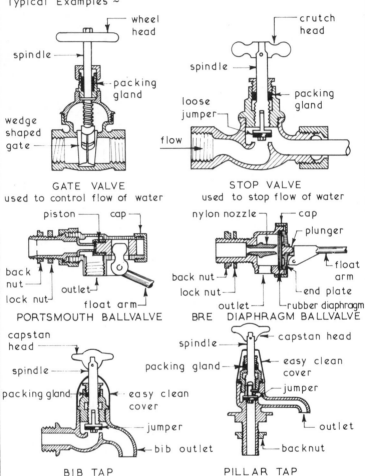

GATE VALVE
used to control flow of water

STOP VALVE
used to stop flow of water

PORTSMOUTH BALLVALVE

BRE DIAPHRAGM BALLVALVE

BIB TAP
horizontal inlet - used over sinks and for hose pipe outlets

PILLAR TAP
vertical inlet - used in conjunction with fittings.

Fig 82   Hot and cold water installations — flow controls

Cisterns ~ these are fixed containers used for storing water at atmospheric pressure. The inflow of water is controlled by a ballvalve which is adjusted to shut off the water supply when it has reached the designed level within the cistern. The capacity of the cistern depends on the draw off demand and whether the cistern feeds both hot and cold water systems. Domestic cold water cisterns should be placed at least 750mm away from an external wall or roof surface and in such a position that it can be inspected, cleaned and maintained. A minimum clear space of 300mm is required over the cistern for ballvalve maintenance. An overflow or warning pipe of not less than 19mm diameter must be fitted to fall away in level to discharge in a conspicuous position. All draw off pipes must be fitted with a stop valve positioned as near to the cistern as possible.

Cisterns are available in a variety of sizes and materials such as galvanised mild steel (BS 417), asbestos cement (BS 2777), plastic moulded (BS 4213) and glass fibre. If the cistern and its associated pipework are to be housed in a cold area such as a roof they should be insulated against freezing.

Typical Details ~

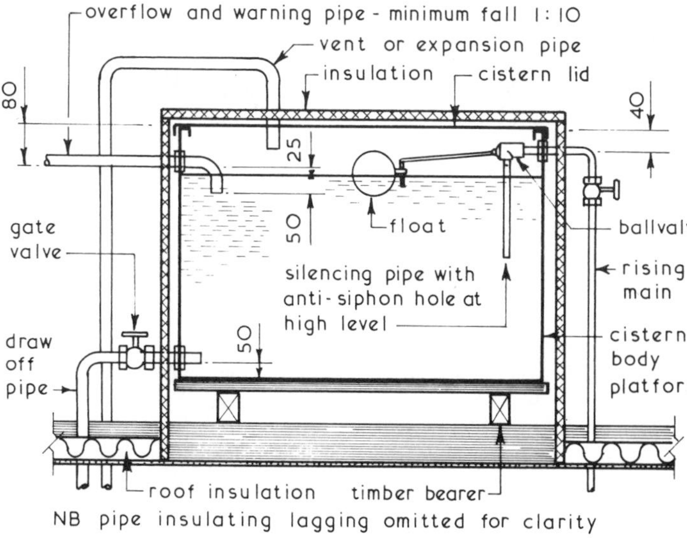

Fig 83   Hot and cold water installations — cisterns

Indirect Hot Water Cylinders ~ these cylinders are a form of heat exchanger where the primary circuit of hot water from the boiler flows through a coil or annulus within the storage vessel and transfers the heat to the water stored within. An alternative hot water cylinder for small installations is the single feed or 'Primatic' cylinder which is self venting and relies on two air locks to separate the primary water from the secondary water. This form of cylinder is connected to pipework in the same manner as for a direct system (see Fig. 80) and therefore gives savings in both pipework and fittings. Indirect cylinders usually conform to the recommendations of BS1565 (galvanised mild steel) or BS1566 (copper).

Typical Examples ~

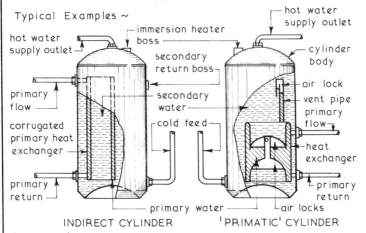

INDIRECT CYLINDER   'PRIMATIC' CYLINDER

'Primatic' Cylinders ~

1. Cylinder is filled in the normal way and the primary system is filled via the heat exchanger, as the initial filling continues air locks are formed in the upper and lower chambers of the heat exchanger and in the vent pipe.

2. The two air locks in the heat exchanger are permanently maintained and are self-recuperating in operation. These air locks isolate the primary water from the secondary water almost as effectively as a mechanical barrier.

3. The expansion volume of total primary water at a flow temperature of 82° C is approximately $1/25$ and is accommodated in the upper expansion chamber by displacing air into the lower chamber, upon contraction reverse occurs.

Fig 84   Indirect hot water cylinders

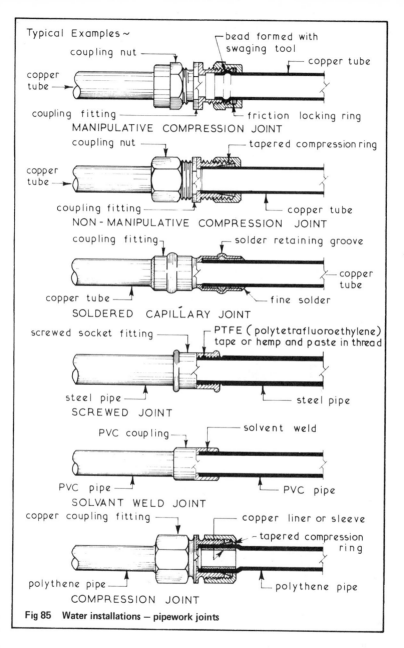

Typical Examples ~

bead formed with swaging tool

coupling nut

copper tube

coupling fitting

copper tube

friction locking ring

MANIPULATIVE COMPRESSION JOINT

coupling nut

tapered compression ring

copper tube

coupling fitting

copper tube

NON-MANIPULATIVE COMPRESSION JOINT

coupling fitting

solder retaining groove

copper tube

copper tube

copper tube

fine solder

SOLDERED CAPILLARY JOINT

screwed socket fitting

PTFE (polytetrafluoroethylene) tape or hemp and paste in thread

steel pipe

steel pipe

SCREWED JOINT

PVC coupling

solvent weld

PVC pipe

PVC pipe

SOLVANT WELD JOINT

copper coupling fitting

copper liner or sleeve

tapered compression ring

polythene pipe

polythene pipe

COMPRESSION JOINT

Fig 85   Water installations — pipework joints

105

Typical Examples ~

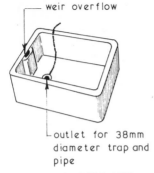

— weir overflow

└ outlet for 38mm
diameter trap and
pipe

BELFAST PATTERN SINK

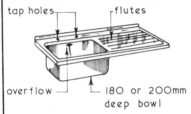

tap holes ─┐   ┌─ flutes

overflow ─┘   └─ 180 or 200mm
deep bowl

SINGLE DRAINER STAINLESS STEEL SINK

Fireclay Sinks (BS 1206) -
these are white glazed sinks
and are available in a wide
range of sizes from 460 x
380 x 200 deep up to 1220 x
610 x 305 deep and can be
obtained with an integral
drainer. They should be fixed
at a height between 850 and
920mm and supported by legs,
cantilever brackets or dwarf
brick walls.

Metal Sinks (BS 1244) -
these can be made of enamelled
cast iron, enamelled pressed
steel or stainless steel with
single or double drainers in
sizes ranging from 1070 x
460 to 1600 x 530 supported
on cantilever brackets or sink
cupboards.

Ceramic Wash Basins (BS 1188)

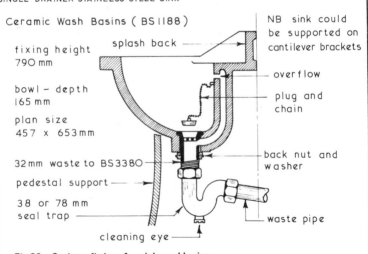

fixing height
790 mm

bowl — depth
165 mm

plan size
457 x 653mm

32mm waste to BS 3380

pedestal support ─→

38 or 78 mm
seal trap

cleaning eye ─────

splash back ─────

NB sink could
be supported on
cantilever brackets

─ overflow

─ plug and
chain

─ back nut and
washer

─ waste pipe

Fig 86   Sanitary fittings 1 — sinks and basins

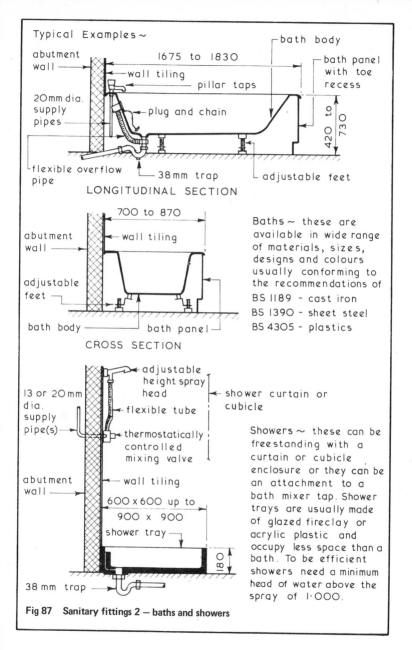

Typical Examples ~

LONGITUDINAL SECTION

- bath body
- abutment wall
- wall tiling
- pillar taps
- bath panel with toe recess
- 1675 to 1830
- 20mm dia. supply pipes
- plug and chain
- 420 to 730
- flexible overflow pipe
- 38mm trap
- adjustable feet

CROSS SECTION

- 700 to 870
- abutment wall
- wall tiling
- adjustable feet
- bath body
- bath panel

Baths ~ these are available in wide range of materials, sizes, designs and colours usually conforming to the recommendations of

BS 1189 - cast iron
BS 1390 - sheet steel
BS 4305 - plastics

- 13 or 20 mm dia. supply pipe(s)
- adjustable height spray head
- shower curtain or cubicle
- flexible tube
- thermostatically controlled mixing valve
- abutment wall
- wall tiling
- 600 x 600 up to 900 x 900
- shower tray
- 180
- 38 mm trap

Showers ~ these can be freestanding with a curtain or cubicle enclosure or they can be an attachment to a bath mixer tap. Shower trays are usually made of glazed fireclay or acrylic plastic and occupy less space than a bath. To be efficient showers need a minimum head of water above the spray of 1·000.

Fig 87 Sanitary fittings 2 — baths and showers

107

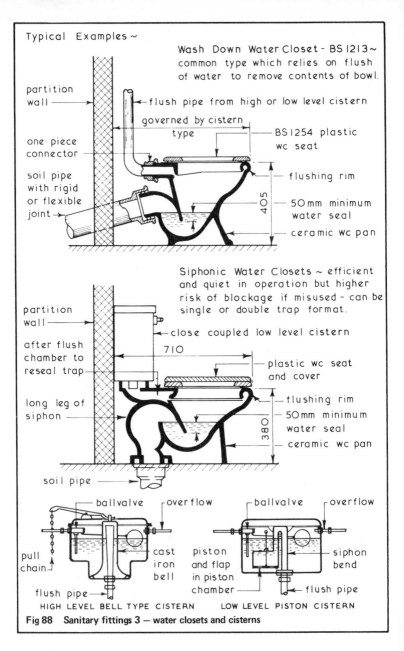

Typical Examples ~

Wash Down Water Closet - BS 1213 ~ common type which relies on flush of water to remove contents of bowl.

partition wall

flush pipe from high or low level cistern governed by cistern type

BS 1254 plastic wc seat

one piece connector

soil pipe with rigid or flexible joint →

flushing rim

405

50 mm minimum water seal

ceramic wc pan

Siphonic Water Closets ~ efficient and quiet in operation but higher risk of blockage if misused - can be single or double trap format.

partition wall

close coupled low level cistern

after flush chamber to reseal trap

710

plastic wc seat and cover

long leg of siphon

flushing rim

380

50 mm minimum water seal

ceramic wc pan

soil pipe →

ballvalve   overflow

pull chain

cast iron bell

flush pipe →

HIGH LEVEL BELL TYPE CISTERN

ballvalve   overflow

piston and flap in piston chamber

siphon bend

flush pipe

LOW LEVEL PISTON CISTERN

Fig 88   Sanitary fittings 3 — water closets and cisterns

108

Single Stack System ~ method developed by the Building Research Establishment to eliminate the need for ventilating pipework to maintain the water seals in traps to sanitary fittings. The slope and distances of the branch connections must be kept within the design limitations given below. This system is only possible when the sanitary appliances are closely grouped around the discharge stack.

Typical Details ~

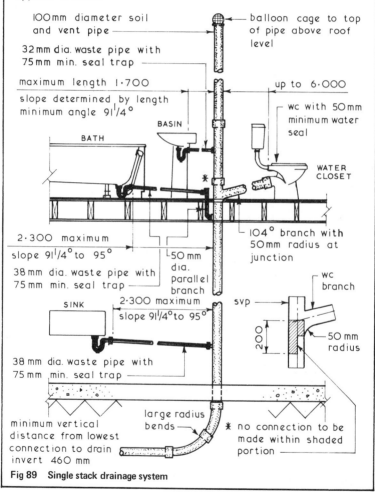

100 mm diameter soil and vent pipe

balloon cage to top of pipe above roof level

32 mm dia. waste pipe with 75 mm min. seal trap

maximum length 1·700

up to 6·000

slope determined by length minimum angle 91¼°

wc with 50 mm minimum water seal

BASIN

BATH

WATER CLOSET

2·300 maximum

slope 91¼° to 95°

104° branch with 50 mm radius at junction

38 mm dia. waste pipe with 75 mm min. seal trap

50 mm dia. parallel branch

SINK

2·300 maximum

slope 91¼° to 95°

wc branch

svp

200

50 mm radius

38 mm dia. waste pipe with 75 mm min. seal trap

minimum vertical distance from lowest connection to drain invert 460 mm

large radius bends

✳ no connection to be made within shaded portion

Fig 89   Single stack drainage system

Inspection Chambers ~ these are sometimes called manholes and provide a means of access to a drainage system for the purpose of inspection and maintenance. They should be positioned in accordance with the requirements of Part N of the Building Regulations. In domestic work inspection chambers can be of brick, precast concrete or preformed in plastic for use with patent drainage systems. The size of an inspection chamber depends on depth to invert level, size and number of branch drains to be accommodated within the chamber — guidance to sizing is given in CP 301.

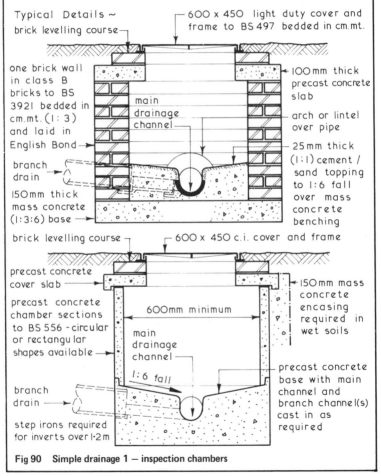

Typical Details ~
brick levelling course

600 x 450 light duty cover and frame to BS 497 bedded in cm.mt.

one brick wall in class B bricks to BS 3921 bedded in cm.mt. (1:3) and laid in English Bond

100mm thick precast concrete slab

main drainage channel

arch or lintel over pipe

branch drain

25mm thick (1:1) cement / sand topping to 1:6 fall over mass concrete benching

150mm thick mass concrete (1:3:6) base

brick levelling course

600 x 450 c.i. cover and frame

precast concrete cover slab

150mm mass concrete encasing required in wet soils

precast concrete chamber sections to BS 556 - circular or rectangular shapes available

600mm minimum

main drainage channel

1:6 fall

precast concrete base with main channel and branch channel(s) cast in as required

branch drain

step irons required for inverts over 1·2m

Fig 90   Simple drainage 1 — inspection chambers

110

Excavations ~ drains are laid in trenches which are set out, excavated and supported in a similar manner to foundation trenches except for the base of the trench which is cut to the required gradient or fall.

Typical Detail ~

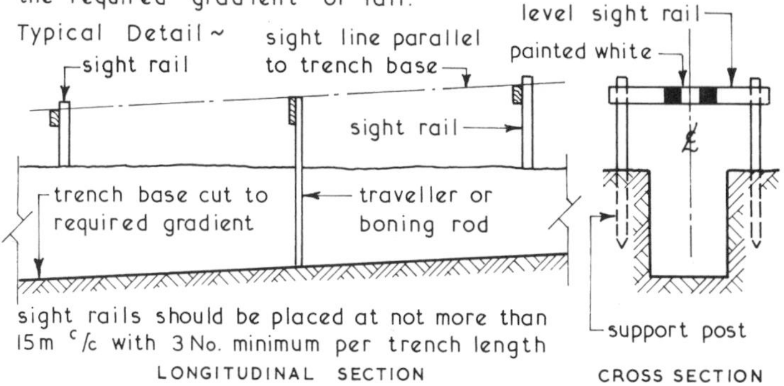

sight line parallel to trench base

level sight rail
painted white
sight rail
sight rail
trench base cut to required gradient
traveller or boning rod
support post

sight rails should be placed at not more than 15m °/c with 3 No. minimum per trench length

LONGITUDINAL SECTION   CROSS SECTION

Joints ~ these must be watertight under all working and movement conditions and this can be achieved by using rigid and flexible joints in conjunction with the appropriate bedding.

Typical Joint Details ~

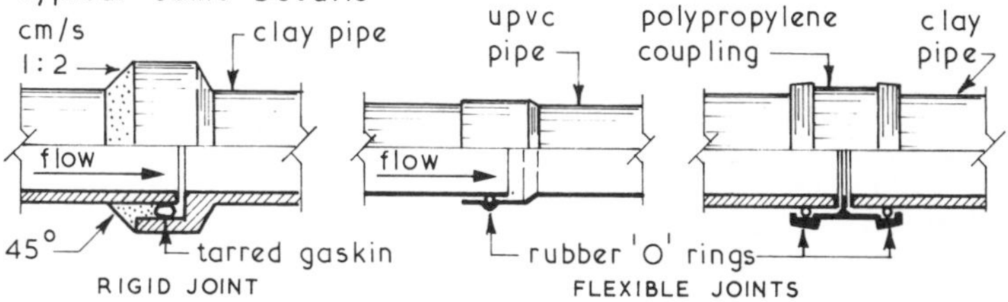

cm/s 1:2
clay pipe
upvc pipe
polypropylene coupling
clay pipe
flow
flow
45°
tarred gaskin
rubber 'O' rings
RIGID JOINT   FLEXIBLE JOINTS

Typical Bedding Details ~

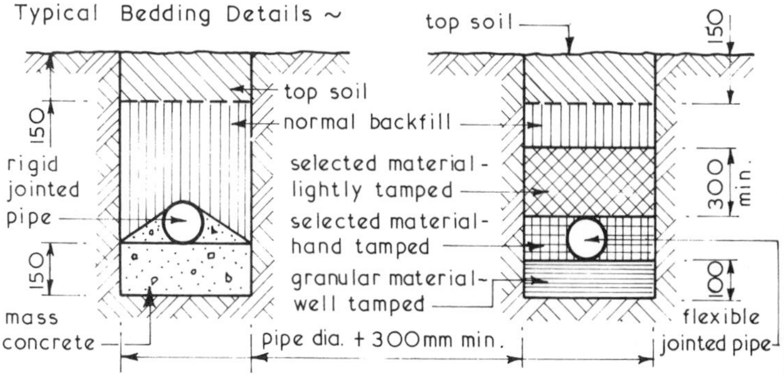

top soil
150
rigid jointed pipe
150
mass concrete
top soil
normal backfill
selected material - lightly tamped
selected material - hand tamped
granular material - well tamped
pipe dia. + 300mm min.
300 min.
100
flexible jointed pipe

Fig 91   Simple drainage 2 — jointing and bedding pipes

111

Highway Drainage ~ the stability of a highway or road relies on two factors —

1. Strength and durability of upper surface.
2. Strength and durability of subgrade which is the subsoil on which the highway construction is laid.

The above can be adversely affected by water therefore it may be necessary to install two drainage systems. One system (subsoil drainage) to reduce the flow of subsoil water through the subgrade under the highway construction and a system of surface water drainage.

Typical Highway Subsoil Drainage Methods ~

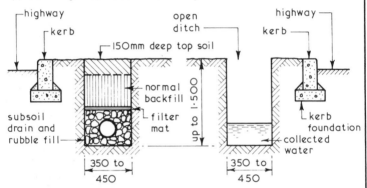

Subsoil Drain - acts as a cut off drain and can be formed using perforated or porous drain pipes. If filled with rubble only it is usually called a French or rubble drain.

Open Ditch - acts as a cut off drain and could also be used to collect surface water discharged from a rural road where there is no raised kerb or surface water drains.

Surface Water Drainage Systems ~

Fig 92   Highway drainage

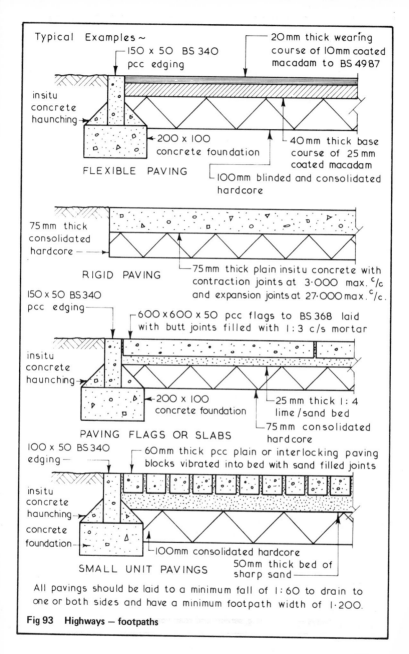

Typical Examples ~

150 × 50 BS 340 pcc edging

20mm thick wearing course of 10mm coated macadam to BS 4987

insitu concrete haunching

200 × 100 concrete foundation

40mm thick base course of 25mm coated macadam

100mm blinded and consolidated hardcore

FLEXIBLE PAVING

75mm thick consolidated hardcore

75mm thick plain insitu concrete with contraction joints at 3·000 max. c/c and expansion joints at 27·000 max. c/c.

RIGID PAVING

150 × 50 BS 340 pcc edging

600 × 600 × 50 pcc flags to BS 368 laid with butt joints filled with 1:3 c/s mortar

insitu concrete haunching

200 × 100 concrete foundation

25mm thick 1:4 lime/sand bed

75mm consolidated hardcore

PAVING FLAGS OR SLABS

100 × 50 BS 340 edging

60mm thick pcc plain or interlocking paving blocks vibrated into bed with sand filled joints

insitu concrete haunching

concrete foundation

100mm consolidated hardcore

50mm thick bed of sharp sand

SMALL UNIT PAVINGS

All pavings should be laid to a minimum fall of 1:60 to drain to one or both sides and have a minimum footpath width of 1·200.

Fig 93 Highways — footpaths

113

Landscaping ~ in the context of building works this would involve reinstatement of the site as a preparation to the landscaping in the form of lawns, paths, pavings, flower and shrub beds and tree planting. The actual planning, lawn laying and planting activities are normally undertaken by a landscape subcontractor. The main contractor's work would involve clearing away all waste and unwanted materials, breaking up and levelling surface areas, removing all unwanted vegetation, preparing the subsoil for and spreading topsoil to a depth of at least 150 mm.

Services ~ the actual position and laying of services is the responsibility of the various service boards and undertakings. The best method is to use the common trench approach, avoid as far as practicable laying services under the highway.

Typical Common Trench Details ~

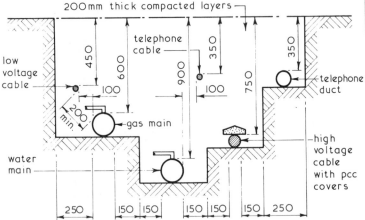

Road Signs ~ these can range from markings painted on roads to define traffic lanes, rights of way and warnings of hazards to signs mounted above the road level to give information, warning or directives, the latter being obligatory.

Typical Examples ~

INFORMATION   WARNING - ROAD WORKS   DIRECTIVE - NO LEFT TURN

Fig 94   Highways — landscaping, services and road signs

Simple Small Span Bridges ~ these can be constructed from any structural material such as timber, insitu reinforced concrete, precast concrete, steel and aluminium alloys or any combination of these materials but whichever is chosen the bridge will be designed to the same basic principles and will usually be composed of the same basic members.

Typical Example ~

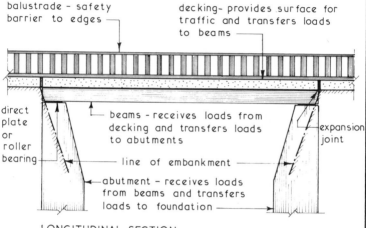

balustrade - safety barrier to edges

decking - provides surface for traffic and transfers loads to beams

direct plate or roller bearing

beams - receives loads from decking and transfers loads to abutments

line of embankment

expansion joint

abutment - receives loads from beams and transfers loads to foundation

LONGITUDINAL SECTION

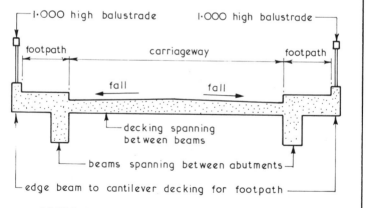

1·000 high balustrade

1·000 high balustrade

footpath

carriageway

footpath

fall

fall

decking spanning between beams

beams spanning between abutments

edge beam to cantilever decking for footpath

CROSS SECTION

Fig 95   Simple small span bridges

Index